THE GREAT INDIAN PHONE BOOK

ASSA DORON and ROBIN JEFFREY

The Great Indian Phone Book

How the Cheap Cell Phone Changes Business, Politics, and Daily Life

NEW HANOVER COUNTY
PUBLIC LIBRARY
201 CHESTNUT STREET
WILMINGTON, NC 28401

HARVARD UNIVERSITY PRESS
Cambridge, Massachusetts
2013

Copyright © Assa Doron and Robin Jeffrey, 2013
All rights reserved
Printed in the United States of America

First published in the United Kingdom in 2013 by
C. Hurst & Co (Publishers) Ltd.,
41 Great Russell Street, London, WC1B 3PL

First Harvard University Press edition, 2013

Library of Congress Cataloging-in-Publication Data

Doron, Assa.
The great Indian phone book : how the cheap cell phone changes business, politics, and daily life / Assa Doron and Robin Jeffrey.
p. cm.
Includes bibliographical references and index.
ISBN 978-0-674-07268-8 (cloth : alk. paper)
1. Cell phones—Social aspects—India.   I. Jeffrey, Robin.   II. Title.
HE9715.I4D67 2013
384.5'350954—dc23   2012040090

CONTENTS

Preface	ix
Glossary	xv
Abbreviations	xvii
List of Maps, Illustrations, Figures and Tables	xxi
Acknowledgements	xxvii
Radio Frequency and Mobile Phones	xxxi
Introduction: 'So Uncanny and Out of Place'	1
In India	2
In the world	9
In conclusion	13

PART ONE
CONTROLLING

1. Controlling Communication	19
Horses, runners and rulers	20
Untying communication	32
2. Celling India	39
Act I: '… Within a fortnight …'	41
Act II: Sidelining the referee	47
Act III: Bread, clothing, shelter—and a mobile	53
Act IV: Schools for scandal	57

PART TWO
CONNECTING

3. Missionaries of the Mobile	65
Man's best friend	66

CONTENTS

Talk time—small, medium, large	70
The art of retail	75
4. Mechanics of the Mobile	89
People	90
Factory workers	90
Tower walas	94
Mistriis	97
Trainers and trainees	100
Process	104
The Care Centre	104

PART THREE
CONSUMING

5. For Business	115
On the sea …	118
Around the globe …	121
At the bank …	123
On the river …	131
On the farm …	137
Empowering, ensnaring or just chatting …?	141
6. For Politics	143
'Smart mobs' in the world	144
'Smart organisations' in India	146
Limits, lessons and possibilities	158
7. For Women and Households	165
Who will guard the mobile?	170
The household mobile	172
Ownership and property	176
Romance, marriage and the mobile	178
8. For 'Wrongdoing': 'Waywardness' to Terror	185
'Waywardness'	187
Pornography	191
Crime	197
Scandal and surveillance	202
Espionage and terror	205
Conclusion: 'It's the autonomy, stupid'	209
Health	211
Mobile waste	213

CONTENTS

Social networks 215
Language and media 218
Politics and governance 220

Notes 225
Bibliography 265
Index 281

PREFACE

'Who did you write this book for?' a publisher asked us. 'Ourselves', we replied with as close to one voice as possible, given that Doron was in Australia and Jeffrey in Singapore. But it was true: this is the book we both were looking for in 2008 when the puzzle—the miracle, the in-your-faced-ness—of Indian mobile telephony began to strike us every day. Each of us was teased by memorable first encounters with Indian telephones. For Jeffrey, it was the marriage of two Canadians in New Delhi on a sunny Saturday afternoon in December 1967. After the ceremony, a reception was held in the grounds of a grand old New Delhi bungalow, then occupied by a senior member of the Canadian High Commission. From mid-afternoon, attempts were made to telephone the parents of the bride and groom in Toronto. The project came to resemble a disaster-rescue saga. Occasionally, the person placing the call—we took it in shifts—would break into the conviviality with an announcement: 'We nearly got through that time. I think I heard someone pick up ... but it may have been another operator ... Our operator says perhaps in an hour'. Darkness fell, eventually the party broke up, and the call was never completed. That experience was one of the few times Jeffrey had any connection with a telephone during two years in India between 1967 and 1969. The school where he taught in Chandigarh was believed to have a phone, rumoured to be in the principal's office, but no teacher had ever heard it ring or heard that it had been used. No one Jeffrey knew had a phone or thought of using one. What, after all, were bicycles, social visits and Indian Post and Telegraphs (IP&T) for?

PREFACE

For Doron, arriving in India twenty-five years later, an encounter with a telephone was steeped in anxiety and anticipation. As a backpacker in the early 1990s, he soon realised he had to have a plan prior to entering one of the yellow-painted STD booths—the 'PCOs' or Public Call Offices—on the street. Preparation involved making sure that all important information was communicated to those at the other end of the line in the most succinct and powerful manner. This was vital because the line was often cut and costs were high: a day's sustenance—Rs 300—was the cost of a minimum three-minute call.[1] Doron anxiously watched the red-digits displaying the mounting seconds on the electronic clock as he screamed into the mouthpiece in the hope that his voice would carry over the radiowaves to faraway lands. One could easily blow a whole day's budget on a single botched call. Unfortunately, all too often the phone conversations in those bright yellow booths ended with an argument over the price and quality of the call. The experience could be unpleasant. For Doron, visiting the local post office to send postcards and sift leisurely through the post-restante mail was the preferred way to communicate with family and friends.

In the first decade of the twenty-first century, all this changed. The take-up of cell phones from 2004 was rocket-like. (We use 'cell phone' and 'mobile phone' interchangeably). By 2012, mobile-phone subscribers in India exceeded 900 million out of a total population of 1,220 million (1.22 billion). On those numbers, three out of every four Indians, from kids to octogenarians, had a mobile phone. And even if we don't take the 900 million figure too seriously (there is a lot of double-counting), and even when we recognise that phones are much less common among rural poor people than urban rich ones, it means that close to half of Indians almost certainly owned a phone. Doron and Jeffrey were like millions of others who came to marvel at daily encounters with these facts and to add to their story-telling repertoire tales that began: 'The rickshaw guy the other night ... he was pedalling me home and his phone rang! He stopped and told me to get out—he couldn't take me any farther. His wife wanted him to do another job'.

Though we had both used mobile phones in Australia and elsewhere since the mid-1990s, we realised we knew little about them—how they worked, how and where they were made, and

PREFACE

who made money from them and how. Such questions provided the motivation for this book—an attempt to piece together the jigsaw of how cell phones came to India and what their impact has been.

We were also struck by the fact that so many individuals we met carried a mobile regardless of whether they were boatmen or high officials. In the West, the smart-phone revolution was beginning to make its mark, but up to then, mobile phones were just another telephone, often associated with 'work', rather than 'play'. In India, for millions who never had the luxury and opportunity to communicate through a household fixed-line, the arrival of the cheap cell phone was a revolution, and everyone wanted to have at least one in the family—usually the men, but increasingly women too.

Each of the eight chapters of this book is worth a book in itself, and each is imperfect in ways that we as the authors know too well. But, so far as we can see, there is no book about cell phones—not just in India but worldwide—quite like this one. It's an attempt to map the mobile-phone food-chain from the Killer Whales at one end to the Small Fry at the other. We've divided the book into three sections. The first is called *Controlling*. Its two chapters examine how powerful people struggle to control information, beginning with the sub-continent's Mughal rulers 500 years ago but quickly moving to Radio Frequency (RF) spectrum, big business, bureaucrats and politicians today. (There is a note about RF on pp. xxxi-xxxii). The second section is called *Connecting*. It aims to understand how in less than ten years mobile phones, and the technology to support them, found their way into the hands of hundreds of millions of people in a vast country of dispersed population, low literacy and extensive poverty. The third section of the book constitutes half its content. We call it *Consuming*. It explores how people in India use cell phones—in business, politics, domestic life and crime.

We had a few advantages when we set ourselves the too-ambitious task of trying to answer our own questions about cell phones. Between the two of us, we have been powering our intellectual lives off the unpredictable electricity of India for more than sixty years. We also had a range of complementary techniques and skills. Jeffrey was trained as a historian but worked most of his academic life in a university department called 'Politics'.

PREFACE

Doron is an anthropologist. Jeffrey's white hair and forty years of friendships in India sometimes helped to make connections and organise interviews. Doron's anthropologist's affability, solid Hindi and myriad connections in Banaras generated revealing conversations with people in all walks of mobile-phone life. He did what anthropologists do: ethnographic fieldwork, based on observation and interviews. He aimed to de-familiarise what seemed a banal topic: mobile phones. These new, small artefacts, which changed the lives of millions and saturated the Indian landscape, had quickly become commonplace and taken-for-granted—an extension of the self—a prosthetic. Often, we had to remind people that the mobile phone was a recent phenomenon whose entry into their lives was worth considering. 'How', we would ask, 'did you do X before you had a phone?' A pause. 'Oh, we didn't of course'.

Geographically, Jeffrey had friends and interests in Kerala, Punjab and New Delhi. Doron had lived and worked in Uttar Pradesh, Rajasthan and Goa. In researching this book, we travelled from Thiruvananthapuram to Shimla, and Lucknow to Mumbai, on a number of visits of varying durations. The many debts we incurred during those travels are recorded in the Acknowledgements later in the book. We had relatively formal interviews with more than sixty people directly connected with aspects of mobile telephony, and we enjoyed informal conversations with scores of others. It was difficult to talk to us in India without being interrogated about cell phones and what they meant for the person who had fallen into our inquisitive clutches. We read everything we could find on the spread of mobile phones elsewhere in the world and came to admire the work of analysts like Manuel Castells, Jonathan Donner, Heather Horst, Daniel Miller, James Katz, Rich Ling, Howard Rheingold and others. Those debts are recorded in notes and comments in the following chapters. We culled major newspapers and periodicals and drew on government documents and the publications of businesses and non-government organisations. The Web makes some of this work quicker and easier than it would have been twenty years ago. But sitting at a computer screen is a poor substitute for visiting a crowded training institute for aspiring technicians on the outskirts of Banaras or trailing behind an engineer as he walks through a rubber estate in Kerala to inspect the air-conditioning unit that cools one of his transmission towers.

PREFACE

The writing of *The Great Indian Phone Book* proved as joint a project as one can imagine 'joint authorship' being. As the structure of the book evolved, one of us took responsibility for writing the first draft of particular chapters. Doron wrote first drafts of chapters dealing with the social uses of mobile phones; Jeffrey started chapters on history, politics and marketing. The other author then took the first draft and objected, deleted, edited and added. It was a bit like planting a garden or playing pingpong. One author planted, the other weeded, added his own favourites and turned the plot back to the originator to go through the weeding-and-planting process again. The pingpong ball went back and forth across the net many times, and we infected each other with daily doses of enthusiasm.

We aimed to write a book that would hold up its head as both sound scholarship and engaging reading. Our potential readers were us: curious people, eager for understanding and intolerant of jargon. We have tried to make the book accessible. Mobile telephony produces an ABC of acronyms; we have tried to use them sparingly and have provided a list of Abbreviations for quick reference. Similarly, the Glossary contains words that may be unfamiliar to people who are not electronics engineers as well as to people who are not Indians or Indophiles. The two Maps provide a reminder of India's size and diversity and identify the 'circles' (geographic units) into which the Government of India has divided, and leased out, the right to use Radio Frequency for mobile telecommunications.

We have not used diacritical marks for Indian words. When transliterating extended passages from an Indian language, we show long vowels by repeating the letter (for example, 'aa'), except at the end of words, in which case they are simply 'a'. For words that are in wide use, we follow common spellings. A 'wala' remains a 'wala' and the god Ganesh stays 'Ganesh', though a more pedantic transliteration would have 'vaala' and 'Gaṇeśa'. Official spellings and direct quotations appear as they do in the originals.

For the title of the book, we are indebted to Michael Dwyer. We were keen on 'Celling India' as a title, but people pointed out that 'cell phone' was not a term used universally. We toyed with alternatives, including variations of the overused 'revolution', before Michael suggested the teasing title we adopted. To his ear and ours, a phrase beginning, 'the great Indian' would once have

PREFACE

ended in the words, 'rope-trick'—the old Orientalist myth of the Indian magician who could make a rope stand straight in the air, then climb it and disappear. That's the kind of magic that mobile phones offer. And 'phone book' is a term everyone has heard (but children born after the year 2000 are unlikely ever to see). 'Phone book' suggests a reference, something that contains a lot of information and, in its way, is complete. So it became 'the great Indian phone book'.

Early attempts to come to terms with issues raised in this book appeared in journals and magazines. We thank the editors of the *Journal of Asian Studies*, *South Asian History and Culture*, *Pacific Affairs* and *The Asia Pacific Journal of Anthropology*, as well as various media outlets, for publishing the earlier work and enabling us to develop ideas that we explore here.

In the Acknowledgements, we thank the host of people whose kindness, patience and knowledge have helped us so generously. Here, we thank the institutions that have supported us: the Australian Research Council, the College of Asia and the Pacific at the Australian National University, and the Institute of South Asian Studies and the Asia Research Institute at the National University of Singapore, where Jeffrey benefited from a delightful and stimulating environment for three years. Domestically, Jeffrey is, as usual and forever, indebted to Lesley for tolerating constant chatter about India, now made even more glorious by references to 'RF', 'fibre optic' and various 'Gs' from '2' to '4'. For Doron, his family and especially Udi, Harry, Raya and Guy provided invaluable intellectual and emotional support. Minnie and the kids (Itai and Tomer) were a constant reminder of the most important things in life; they shared the passion and curiosity for India, with many memorable moments on north Indian trains and flying kites from the boats of Banaras.

Robin Jeffrey　　　　　　　　　　　　　　　　　　Assa Doron
Singapore　　　　　　　　　　　　　　　　　　　*Canberra*
October 2012

GLOSSARY

aam aadmii	common man
bahu	daughter-in-law
barakat	a blessing; good fortune, charismatic power
bhaaichaara samiti	a brotherhood committee set up to bring different groups together
bhajan	devotional song
Bluetooth	set of standards for low-power, short-range wireless communication between devices like cell phones and computers; slower and more limited than Wifi
Code Division Multiple Access (CDMA)	set of standards for transmitting radio frequency signals; adopted in USA and by Tata, Reliance and BSNL in India
crore	ten million
Dalit	term denoting 'ex'-Untouchable castes in India
darshan	a sacred vision; opportunity to be blessed by exposure to a revered person or object
dhobi	washerman or (less likely) washerwoman
garibi hatao	banish poverty
ghat	the bank of a river, or side of a tank, and the steps leading down to the water
ghatwar	a owner of a boat on the Ganga at Banaras
Global System for Mobile Communications (GSM)	set of standards for transmitting radio frequency signals, developed in Europe and widely used in the rest of the world including India

xv

GLOSSARY

gupshup	gossip
kabaadiwala	a waste merchant; someone who collects cast-off material and recycles it
kaccha	impermanent; not built to last; faulty
keitai	Japanese term for a cell phone
kotwal	a police official in Mughal times
lakh	100,000
mallahi	a rower of boats on the Ganga
mistrii	a skilled artisan
paan	a mixture of betel, lime, spices and sometimes tobacco, wrapped in a green leaf and chewed as a mild recreational addiction
roti, kapda, makaan aur mobile	bread, clothing, shelter and mobile phone
saambaar	fragrant vegetable curry
Subscriber Identity Module (SIM)	thumb-nail-sized card, stamped with digital code and inserted into a GSM phone to give it a unique identity and allow it to connect to a network
wala, wali	someone (male, female) who does something; e.g. a 'Mobile Wali'—a woman with a mobile phone
Wifi	set of technical standards established by the Institute of Electrical and Electronics Engineers (IEEE) that allows transmission of Radio Frequency signals carrying a lot of data quickly over short distances
yaatra	a journey

ABBREVIATIONS

2G, 3G, 4G	2nd, 3rd and 4th generation standards of technology for use of radio frequency to convey data
ARPU	average revenue per user
ASHA	Accredited Social Health Activists
ATM	automatic teller machine
BAMCEF	Backward and Minority Communities Employees' Federation
BDO	Block Development Officer
BJP	Bharatiya Janata Party
BOP	bottom of the pyramid
BPL	below the poverty line
BSNL	Bharat Sanchar Nigam Ltd—the government telecom company for all India except Mumbai and Delhi
BSP	Bahujan Samaj Party
CAG	Comptroller and Auditor-General
CBI	Central Bureau of Intelligence
CCTV	closed-circuit television
CD	compact disc
C-DOT	Centre for Development of Telematics
COAI	Cellular Operators Association of India
CDMA	code division multiple access
DAE	Dhirubhai Ambani Entrepreneur
DoT	Department of Telecommunications
DMK	Dravida Munnetra Kazhagam; political party of Tamil Nadu

ABBREVIATIONS

EDGE	Enhanced Data rates for GSM Evolution technology
EKO	Indian organization aiming to provide mobile-phone-based banking
EMR	electro-magnetic radiation
FMCG	fast moving consumer goods
GHz	giga-hertz (billions of cycles or oscillations per second)
GSM	Global System for Mobile communications (or Groupe Speciale Mondiale)
IAS	Indian Administrative Service
ICICI	Industrial Credit and Investment Corporation of India
ICT	Information and Communications Technologies
IP&T	Indian Post and Telegraph
IKSL	IFFCO Kisan Sanchar Ltd
IIT	Indian Institute of Technology
ITIL	India Telecom Infrastructure Ltd
ITI	Indian Telecom Industries
JNU	Jawaharlal Nehru University
JWT	J. Walter Thompson; advertising firm
MMS	multi-media messaging service
MMT	mobile money transfer
MNC	multi-national corporation
MNP	mobile number portability
MNREGA	Mahatma Gandhi National Rural Employment Guarantee Act
Mpbs	mega-bytes per second
M-PESA	mobile-phone-based banking organization originating in Kenya
MTNL	Mahanagar Telephone Nigam Ltd—the government telecom company for Mumbai and Delhi
NGO	non-government organization
NTP	National Telecom Policy, 1994 and 1999
OBC	Other Backward Classes
OS	operating system
PCO	Public Call Office
RF	radio frequency
RTL	Reliance Telecom Ltd
SC	Scheduled Castes

ABBREVIATIONS

SDP	State Domestic Product
SEZ	Special Economic Zone
SIM	subscriber identity module—the tiny circuit card that goes in a mobile phone
SMS	short message service
ST	Scheduled Tribes
TDMA	time division multiple access
TDSAT	Telecom Dispute Settlement and Appellate Tribunal
TRAI	Telecom Regularly Authority of India
UIAI	Unique Identification Authority of India; also known as Aadhaar
UAPA	Unlawful Activities Prevention Act
UASL	universal access service licence
UP	Uttar Pradesh
USO	universal service obligation
VAS	value added services
VSNL	Videsh Sanchar Nigam Ltd—overseas telecom company, acquired from the government by Tata in 2008
Wifi	radio frequency standards for short-distance data transfer
WLL	wireless in local loop

LIST OF MAPS, ILLUSTRATIONS, FIGURES AND TABLES

MAPS

1. Union of India
   States, state capitals and places mentioned in the text. xxxiii
2. Telecommunications circles
   A, B, C and Metro circles as declared by the Government of India. xxxiv

ILLUSTRATIONS
(Between pages 142 and 143)

Illus. 1. Multi-tasking. Ganga River boatman ferries passengers and checks phone. Banaras ghats in background. July 2012. (Photo: Sachidanand Dixit with thanks).
Illus. 2. Love in a time of SMS. *India Today*'s artist gave 17th-century lovers 21st-century devices. (*India Today*, 14 October 2002, with permission).
Illus. 3. Vanishing species. Yellow-painted Public Call Offices (PCOs) were everywhere but for fewer than 20 years. (Photo: Wikipedia, downloaded July 2012).
Illus. 4. *Swamy's Treatise on Telephone Rules*. The 850-page guide to the mysteries of the government telephone monopoly was a profit-maker as late as 1993. (Photo: Robin Jeffrey).
Illus. 5. Duelling telcos. Vodafone and the government provider BSNL fight for attention in Banaras. October 2009. (Photo: Assa Doron).

LIST OF MAPS, ILLUSTRATIONS, FIGURES AND TABLES

Illus. 6. Cheeka to the rescue. The mobile-phone dog started her advertising career with Hutch and helped move the brand to Vodafone. Cheeka, like the cell-phone service she represents, is always there when you need her or so the ad would like us to believe. (Ogilvy and Mather and Vodafone, with permission and thanks).

Illus. 7. Educating consumers. In one of the many Vodaphone outlets in Banaras, a placard explains to English speakers the wonders of mobile communication—from 'SMS' to '3G' and 'Wifi'. October 2009. (Photo: Assa Doron).

Illus. 8. 'I have the touch', Bollywood superstar Aamir Khan, 'Brand Ambassador' for Samsung, tells buyers. The handwritten notice makes clear to newcomers to capitalism: 'Fixed Prices, No Bargaining'. October 2009. (Photo: Assa Doron).

Illus. 9. The art of retail, Part 1. Ravi's privileged Samsung Mobile Outlet. Banaras, June 2010. (Photo: Assa Doron).

Illus. 10. The art of retail, Part 2. Sales promoters explain features in Ravi's shop. Banaras, June 2010. (Photo: Assa Doron).

Illus. 11. The art of retail, Part 3. Nokia display at Samir's modest shop. Banaras, October 2009. (Photo: Assa Doron).

Illus. 12. Seasonal work. Sumit came to Samir's shop to promote Nokia mobiles prior to the Diwali festival in October 2009. (Photo: Assa Doron).

Illus. 13. 'Make Distance Vanish'. Cheap calls made long-distance romance possible. 50-paise-per-minute (one US cent). October 2009. (Photo: Assa Doron).

Illus. 14. Three towers out of 400,000 across India. North side of Delhi. July 2012. (Photo: Assa Doron).

Illus. 15. Road-side fixer. Lucknow's Hazratganj. June 2010. (Photo: Assa Doron).

Illus. 16. Pavement paraphernalia. Mobile-phone gear includes batteries, chargers, cases and more. Delhi. February 2011. (Photo: Assa Doron).

Illus. 17. Mushrooming industry. Diploma from a mobile-phone repair institute. Banaras. February 2011 (Photo: Assa Doron).

Illus. 18. 'Choreography of consumerism'. Nokia's classy service centres may alienate poor customers. Banaras, October 2009. (Photo: Assa Doron).

LIST OF MAPS, ILLUSTRATIONS, FIGURES AND TABLES

Illus. 19. 'Nokia laaiif tuuls [life tools]. Valuable information—within your reach'. New Delhi. January 2010. (Photo: Assa Doron).

Illus. 20. Tied to the state. Getting a SIM card requires filling out a form, attaching a photograph and providing personal details. (Photo: Assa Doron).

Illus. 21. Sea cells in south India. Kerala fisherman and their phones gained early fame. (Photo: *The Hindu*, 17 May 2012, with permission and thanks).

Illus. 22. Banking comes to a shop near you. An EKO bank outlet (large white signboard, top centre) and Fast Moving Consumer Goods. New Delhi. February 2011. (Photo: Assa Doron).

Illus. 23. Everybody's doing it. From Airtel's smooth middle classes to nicely posed rickshaw pedallers. Chandigarh. May 2009. (Photo: Ajay Varma, Reuters, with permission).

Illus. 24. Communication technology, old style. Kanshi Ram, organizational genius of the Bahujan Samaj Party, on a cycle *yaatra*. (Photo: http://www.ambedkartimes.com/sahib_kanshi_ram.htm. Photographer and original place of publication unknown).

Illus. 25. SMSing to the faithful. Message to Bahujan Samaj Party workers calling on them to celebrate the 75th birthday of their late founder, Kanshi Ram, in 2009. Blue is the party colour—thus 'Blue Salute'. (Photo: Robin Jeffrey, June 2010).

Illus. 26. Communication technology, Mark II. Akhilesh Yadav takes up Kanshi Ram's bicycle and adds a mobile phone for the 2012 election campaign in Uttar Pradesh. (*India Today*, 5 March 2012, with permission).

Illus. 27. Communication technology, Mark III. Tribal women send news items to CGNet Swara. (Photo: Purusottam Singh Thakur and CGNet Swara, with permission and thanks).

Illus. 28. Phoning or broadcasting? Tribal woman sends message to SGNet Swara. (Photo: Purusottam Singh Thakur and CGNet Swara, with permission and thanks).

Illus. 29. Buying vegetables? Unsettling society? Mobile phones force families to make choices. Why didn't she ring her vegetable seller and ask him to deliver? Banaras. October 2009. (Photo: Assa Doron).

Illus. 30. Who owns the mobile? India's leading mobile-phone magazine calls itself *My Mobile* and puts phone-wielding women on its cover. Elsewhere, families agonized about whether women should have phones. (*My Mobile*, Hindi edition, September 2011, with permission and thanks).

Illus. 31. A woman with a phone excited some people. Bhojpuri video clip, *Mobile Wali: Woman with a Mobile Phone*. Singers: Manoj Tiwari, 'Mridul' and Trishna. Publisher: WAVE VCD. (Accessed from YouTube on 23 July 2012. Screen dump by Paul Brugman, Australian National University).

Illus. 32. For some, a phone in a woman's hand was a disturbing accessory. *Mobile Wali: Woman with a Mobile Phone*. Singers: Manoj Tiwari, 'Mridul' and Trishna. Publisher: WAVE VCD. (YouTube, 23 July 2012. Screen dump by Paul Brugman, Australian National University).

Illus. 33. 'The Washerwoman with the Mobile Phone'. Cover of DVD entitled *Mobile Wali Dhobinaya*. Singers: Dinesh Lal Gaundh and Noorjehan. Publisher: GANGA, VCD. (Photo: Paul Brugman, Australian National University).

Illus. 34. Youth market. 'Adult film star' Sunny Leone became brand ambassador for Chaze mobiles. Chaze chased younger buyers with the offer of cheap multimedia phones. (*My Mobile*, 12 July 2012, with permission and thanks).

Illus. 35. Terror and an amulet. A burned victim of the Mumbai bombings of 13 July 2011 clutches a connection to aid and succour—his cell phone. (*Outlook*, 25 July 2011, with permission and thanks).

Illus. 36. Brain tumours, heart attacks, cancer and impotence are some of the things mobile-phone radiation can do to you, according to this advertisement on a men's toilet in New Delhi in July 2012. Prabhatam, the advertiser, offered 'radiation-safe mobile solutions'. (Photo: Assa Doron).

Illus. 37. SIM card throwaways. Discarded telecom goods pose growing challenges. The tiny SIM-circuit portion of

LIST OF MAPS, ILLUSTRATIONS, FIGURES AND TABLES

            these cards has been popped out and inserted in a phone. (Photo: Toxic Link, with permission and thanks).
Illus. 38. This adult-sized mound of waste was being picked over by cottage-industry waste-recyclers. Outskirts of New Delhi, 2012. (Photo: Toxic Link, with permission and thanks).

FIGURES

Fig. 1. Phone subscribers in India, 1998–2012, Wireline and Wireless, in millions.   7

TABLES

Table 1.1. Phone Connections in India, 1947 to 2011.   28
Table 1.2. Items carried by Indian Posts, 2004–05 to 2008–09.   29

ACKNOWLEDGEMENTS

In the Preface, we acknowledged debts to institutions and to the very special family people who put up with us on a limited-warranty, no-refund, no-return basis. Here, we try to thank those who helped us in a great many different and special ways. Some are longtime victims of our writing, consulted regularly for their willingness both to encourage and to warn. Into this category fall Alex Broom, Dipesh Chakrabarty, Nalin Mehta, Kama Maclean, Talis Polis, Philip Taylor and Ira Raja.

Others, like Kishor Dabke, have given us very special help—Kishor, in trying to improve our understanding of Radio Frequency technology and what it means for mobile telephony. He bears no responsibility, however, for errors that remain: they are our fault. Charles Thomson became for us a set of ever-inquisitive ears and eyes roaming north India in the course of his work with EKO and providing us with leads about mobile-phone developments. David Brewster, as a lawyer once involved in telecom contracting, read parts of the manuscript. Philip Lutgendorf read, listened, suggested and sent us photographs. Sonia Sharma and her colleagues at *MyMobile* have never turned us away when we have come calling on them for advice and connections. V. Thiruppugazh is someone from whose guidance we have frequently profited.

On visits to Lucknow and Allahabad, we benefited from the generosity of time and hospitality of M. Aslam (Allahabad University), Vivek Kumar (JNU), Badri Narayan (Pant Institute) and the journalist Sharat Pradhan.

ACKNOWLEDGEMENTS

Friends tolerated a diet of telephone tittle-tattle and were still willing to abet our enthusiasm by putting us in touch with their connections able to help us with our inquiries. Nir Avieli, Frank Conlon, Eva Fisch, Maxine Loynd, Barbara Nelson, Amitendu Palit, Peter Mayer, Ronit Ricci, Jesse Rumsey-Merlan, Ronojoy Sen, Ornit-Shani, Pratima Singh, David Shulman, Kate Sullivan, Matt Wade, Thomas Weber and Tamir Yahav all fall into the category of patient listeners and generous helpers.

Helen Parsons in Canberra helped us shape the manuscript by bringing to bear on it the attentions of a skilful editor. Lee Li Kheng, a delightful cartographer at the National University of Singapore, drew the maps. We were remarkably lucky to encounter two editorial executives who shared our enthusiasm for this project. Michael Dwyer at C. Hurst in London, with whom Jeffrey had worked before, gave the project an adrenalin charge with his quick and keen response to our first approach. And Doron's lucky meeting with Sharmila Sen of Harvard University Press connected us with another inspiring energiser.

Scholars like Amita Baviskar, Greg Bailey, Sara Dickey, Rachel Dwyer, Michael H. Fisher, Peter Friedlander, Hugo Gorringe, Charu Gupta, Dennis McGilvray, S. Narayan, Sudha Pai, Anand Pandian, Arvind Rajagopal, Ursula Rao and Clarinda Still were similarly willing to overlook—and encourage—our phone fetish. They read things, offered things, made suggestions and put us onto new sources.

We benefited greatly from getting to know the participants at a workshop on cell phones that we ran at the Institute of South Asian Studies (ISAS) in Singapore in February 2011, sponsored by ISAS, the Australian Research Council and the Ford Foundation. Those participants were Ang Peng Hwa, Shubranshu Choudhary, Subhashish Gupta, Sunil Mani, Janan Mosazai, Nimmi Rangaswamy, Abhishek Sinha, Deepali Sharma, Sirpa Tenhunen, Zhang Weiyu and Ayesha Zainudeen.

From Doron, special thanks go to Ajay 'Pinku' Pandey whose ongoing assistance in translation, interviews and friendship proved as invaluable as ever. Others who offered their time and assistance in India include Pradeep Manjhi, Nitya Pandey, Deepak Sahani, Rakesh Singh, Ravi Varma, Samir Singh, Sumit Churasia, Dr A. B. Singh, and Vimal Mehra. Jeffrey is indebted to

ACKNOWLEDGEMENTS

T. K. Basu, T. V. Ramachandran, Bobby Sebastian, Gopal Srinivasan, P. Vijayan and T. Willington for connections, directions and instructive conversations. Both of us have benefited from the advice of Anshuman Tiwari. The people we have named so far do not include dozens of Controllers, Connectors and Consumers of the cell phone who talked to us in settings varying from the slick corporate offices of Oberoi Garden City, Mumbai, to the busy stalls of Daal Mandi, Banaras, or palm-shaded tower sites around Kochi. Some of these debts are recorded in the footnotes. Our gratitude to so many patient, informative and entertaining people is immense.

RADIO FREQUENCY AND MOBILE PHONES

A radio wave is an electricity-driven spurt of energy. Think of the ripples a stone makes when it falls into a pond. A big stone makes big ripples—with long gaps between each one; a small stone makes small ripples that are closer together. This is where the term 'radio frequency' comes in—how close together, or how frequent, are the electro-magnetic waves that a transmitter spurts out. Old-fashioned AM radio works at frequencies between 535 and 1700 kilohertz. That means the signal or wave oscillates at between 535,000 and 1.7 million cycles every second. If a radio station is assigned a particular frequency (say, 621 kilohertz), people with receivers tune their receivers to pick up only radio signals bumping along at 621 kHz.

AM radio uses relatively low frequencies because that was what the technology was capable of producing when broadcast radio became possible after the First World War. Advances in technology have brought the ability to transmit at higher frequencies—more oscillations per second—and therefore to pack more information into each time interval because each interval contains a larger number of cycles.

Because the range of Radio Frequency (RF) suitable for economic and technological exploitation is finite, there have been national and international agreements since the 1920s about what sorts of uses, and which organisations, may operate on particular frequencies. For example, 2.402 GHz and 2.480 GHz (that is, 2.402 billion and 2.480 billion cycles per second) are set aside for industrial, scientific and medical devices (ISM). For most of us, these

RADIO FREQUENCY AND MOBILE PHONES

are the frequencies that operate garage doors and Bluetooth. These devices are driven by only tiny amounts of energy so their signal does not travel very far—only 10 or 15 yards or metres—and they therefore do not travel ten blocks away to interfere with someone else's garage door.

Increasing refinement of how you can slice RF spectrum—how you can populate the radio waves with information—led to the first mobile phones and to the continuing refinements since the 1990s.

Because usable RF is limited, and users of mobile phones now number billions, you need to use the same frequencies over and over again in different geographic locations. Signals therefore have to be relatively weak. That's why India needs 400,000 cell phone towers to pick up signals from friendly towers and re-send them to other friendly towers nearby.

That's also why we talk about 'cell' phones. Each telephone tower covers a 'cell', a geographic unit which the telecom company has marked out (usually about 10 square miles or 26 square kilometres). The tower in one cell passes on signals to towers in neighbouring cells, and a phone stays connected to the network even as we travel. (It also means that the network, or the police or criminals, if they want to badly enough, can find us).

The expressions 2G, 3G and 4G stand for second, third and fourth generation technology—the new technologies that allow more electro-magnetic surfboard riders to be put onto a wave. 2G brought voice and SMSes. 3G gives moving pictures and interactive gaming. 4G promises huge increases in speed which in turn will enable more complex content and interactivity.

Since RF is deemed to be a public good of which governments are in charge, governments license chunks of RF to telecom companies in return for large rental fees. Contests to get RF licences on favourable terms lead to wheeling and dealing that can put phones into villages, entrepreneurs into penthouses and politicians into jail. Such possibilities feature in this book.

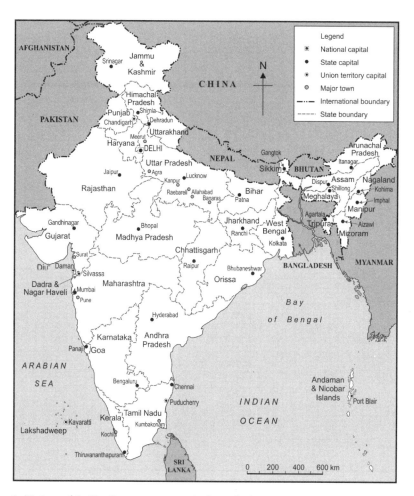

1. Union of India: States, state capitals and places mentioned in the text.

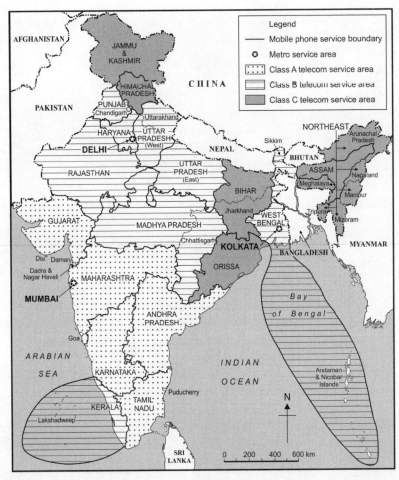

2. Telecommunications circles: A, B, C and Metro circles as declared by the Government of India.

INTRODUCTION

'SO UNCANNY AND OUT OF PLACE'

If we were writing this book as a film script, the opening shot would show a pair of fine slippers lying beside an elegant double bed. Two well-pedicured male feet slip into them. A phone rings. A man's hand picks up a fourth-generation mobile phone, on whose colour screen 'unknown caller' flashes, and the camera pulls back to reveal through a penthouse window the skyline of central Mumbai and the Arabian Sea. The film cuts to battered sandals on a pair of cracked, dusty feet, braced against the deck of what proves to be a large row-boat. The camera frames dawn on the River Ganga, the mist clearing to reveal the ghats at Banaras.[1] (See Illus. 1). Leaning on his oars, a boatload of foreign tourists behind him, a grizzled boatman, fortified against the morning chill by a little alcohol, calls into a battered Nokia, *'Kaun? Mittal Sahib?'* 'Who's that? Is that Mr Mittal?'

The story unfolds ... the boatman, complaining to the well-heeled overseas visitors about inadequate mobile-phone services, has been given by one of them the private number of 'Mr Mittal', founder and chief of a great telecom company, along with the supercilious suggestion that he should let him know of his problem. To the surprise of his patronising passenger, the boatman has no hesitation: he rings Mr Mittal.

The scene is our fantasy. It has never happened; but it could. For the first time, even the poorest, lowest-status people in India can connect to the wealthiest and most highly placed. They can also connect with each other, and they can do so with little hindrance or constraint. Before the mobile phone, such connections were often difficult or impossible, as we describe in the first chapter of this book.² Cheap mobile phones gave poor people a device that improved their chances in a hard world. And in India, because of long-standing discrimination and structures of authority, the mobile has proved even more disruptive than elsewhere.

The focus on footwear in our imagined film script is intentional. The cheap mobile phone is the most disruptive device to hit humanity since shoes. That may seem a tall claim, but consider it. Like shoes, a cell phone allows people to communicate as they were unable to do before. Bare feet limit where you can go. The Vikings who were stealing up to slaughter the sleeping Scots, so the legend has it, gave the game away when their bare feet trod on thistles and they yelped in pain. A young Brahmin on pilgrimage in central India in 1857 echoed the limitations of bare feet: 'my feet ... were oozing blood. It was not possible to heal them, so, often, when the pain got acute, I would melt wax and seal the cracks'.³ In the West, 'until recent times ... to be barefoot meant that all avenues of life were closed'.⁴ The first quality of the male-chauvinist trinity of subjection—'barefoot, pregnant and in the kitchen'—is 'barefoot'. Like shoes, mobile phones have become an item that almost everyone can afford and aspire to.⁵ Like shoes, they go everywhere an individual goes. Like shoes, they show off social class and ideas about fashion. Unlike shoes, mobile phones often get taken to bed. Compare the mobile phone to other devices that human beings have adopted in the past two hundred years. A pencil let you write outside in the rain; a box of matches let you make a fire whenever you needed to; a watch let you be punctual and encouraged others to expect you to be punctual;⁶ a bicycle extended the range you could travel; an automobile let you travel when and where you wished, but needed a lot of money and space to park it. No personal item matched the ubiquity of shoes—until the cheap cell phone.

In India

The cell phone's emergence meant that Indians of every status were able to speak with each other as never before. Those possibilities arose through the widespread dissemination of affordable

INTRODUCTION: 'SO UNCANNY AND OUT OF PLACE'

mobile phones. In February 2012, India had more than 900 million telephone subscribers; 96 per cent were subscribers to mobile phones. Even if we discount this figure by 30 per cent to eliminate duplication and non-active numbers, 600 million subscribers meant a phone for every two Indians, from infants to the aged.[7]

Why are such figures particularly exciting in India? Cell phones have spread throughout the world. A survey in 2012, estimated that there were 6.2 billion mobile subscriptions in the world for a global population of 7 billion.[8] The World Bank pronounced 'the mobile phone network … the biggest "machine" the world has ever seen'.[9] In developed countries, mobile-phone penetration exceeded a hundred per cent. Britain had 76 million mobile phones for 62 million people in 2012. In Africa, the spread of cell phones was also remarkable. Between 2003 and 2008, 'Africa went from having almost no phones to a position where over 100 million Africans' had mobiles.[10] By 2007, Mo Ibrahim, one of Africa's early cell-phone entrepreneurs, could afford to sell his telecom companies, realise US $2 billion and set up a multi-million-dollar foundation to encourage African rulers to hold elections and transfer power voluntarily.[11] The richest man in the world was said to be Carlos Slim, the Mexican telecom tycoon, chairman of Telmex.[12] China made most of the world's mobile-phone components and was estimated to have 900 million mobile-phone users in 2011.[13] Why the fuss about India?

India is both unique and exemplary. Its diversity is unmatched; its caste system has no equivalent for endurance, complexity and malleability; and its experiments with democracy, development and federal government provide unrivalled examples of the potential and the limitations of human endeavour since the industrial revolution. Hierarchy in India was more refined and more deeply embedded in daily life than anywhere else. The lower your status, the less you were entitled to know, to ask and to travel. To be sure, such strictures have changed overtime, faced intense challenge, especially over the last hundred years and were outlawed by the constitution of independent India. But they endure: India remains a caste-conscious society. Mobile phones undermine these strictures. The poor, low-status boatman on the Ganga can conceivably ring India's equivalent of Carlos Slim and can certainly call his friends and contacts and even a town councillor, senior official or Member of Parliament. Those at the top are likely of course to have minions deputed to answer official cell phones. But even if the person who is officially responsible does not answer, a citizen can make other calls: to a media outlet, an opposition party representative, an NGO or an influential relative. Importantly too, 'big people' have several phones, among them their very personal one, which they themselves answer— just as our imaginary Mr Mittal does in our film script. In the past, such connectivity was possible for only a very few people, and as late as 2000, these connections would have been impossible for most Indians. By plugging a large number of previously unconnected people into a system of interactive communication, mobile phones have inaugurated a host of disruptive possibilities.

We have written this book for people like us who wonder every day at head-spinning changes in technology and practices. We puzzle at new terms and ponder how telecommunications in the twenty-first century produced capitalist impresarios in the way that railways and automobiles did in the nineteenth and twentieth centuries. The contribution of *The Great Indian Phone Book* therefore is to try to paint a whole picture—imperfect and incomplete but *whole*—from the corporate captain in the Mumbai penthouse to the weather-beaten oarsman in a boat on the Ganga. Each has been affected by this simple-to-use device.

Other technologies of course prefigured mobile communication and profoundly affected human interaction. The railways, the telegraph and radio all increased the speed with which information could be transferred, and all had mixed implications. They reinforced state control over citizens and contributed to modern imperialism. The telegraph, in particular, made it quick and easy to transmit information across long distances, and governments and great capitalists guarded the technology and used it to mobilise military power, organise commerce and control vast territories. International news agencies, such as Reuters in Britain, Wolff in Germany and Havas in France, transformed telegraphic information into a commodity which they largely controlled.[14]

There are many ways one could study changes brought by the cell phone. We have borrowed a technique from landscape painters to try to capture a whole picture. Our canvas has been divided into three sections—*Controlling*, *Connecting* and *Consuming*. This invites readers to think about how communications were controlled in earlier times—how 'information' or 'intelligence' was valuable and

INTRODUCTION: 'SO UNCANNY AND OUT OF PLACE'

was carefully collected and guarded by those who had it and how rulers feared too much knowledge being bandied about among their subjects. Printing presses in past centuries posed such challenges; but the ability to route Radio Frequency (RF) through a small, personal, cheap device to receive and transmit messages changed the way human beings could interact. It equalised—not wholly, but nevertheless significantly—people's access to information, and began to undermine the epigram that freedom of the press was available only to those who owned one. Mobile phones made everyone a potential publicist. For governments, great corporations and entrepreneurs who would like to be great, the cell phone represented an immense challenge and opportunity.

From 1993 when the technology began to be deployed in India until 2012, the country had ten Ministers of Communication. One of them was convicted of corruption and sent to prison; a second was charged with corruption; a third faced probes that would take years to unravel; a fourth was murdered (though in circumstances not directly related to telecommunications); a fifth was undermined, overruled and rancorously removed.[15] For governments, bureaucrats, regulators and politicians, telecommunications offered a bed of thorny roses, and it is these contests over decision-making and power that we try to understand in the section on *Controlling* which forms Part 1 of this book.

The vast enterprise that bubbled up in the first decade of the twenty-first century generated a cascade of occupations and jobs. There are similarities to the way in which the automobile industry from the 1920s produced new activities and opportunities across the world. A book about the spread of car mechanics in Ghana in West Africa caught this effect well:

> The small-scale industries of the informal sector form a complex matrix with innumerable interconnections. Every enterprise is both a consumer of others' goods and services and a provider to others ...[16]

The mobile phone expanded faster than the automobile. It was cheaper than a car, and many more people were involved in the chain that connected great Controllers to humble Consumers. In Part 2 of this book, we focus on the people who did the *Connecting*. They ranged from the fast-living advertising women and men of Mumbai to small shopkeepers persuaded by their suppliers of Fast Moving Consumer Goods (FMCG) to stock recharge coupons for pre-paid mobile services. In between were technicians who

5

installed transmission equipment; the office workers who found sites and prepared the contracts to install transmission towers (400,000 in 2010);[17] the construction workers and technicians who built and maintained the towers; and the shop owners, repairers and second-hand dealers whose premises varied from slick shopfronts to roadside stalls only slightly more elaborate than those of the repair-walas who once fixed bicycles on the pavement.[18]

But it was the vast numbers of consumers that lent support to the claim that the mobile phone evened the odds—promoted a little more 'equality'—between the powerful and the vulnerable. In the space of less than ten years, beginning from about 2002, India added 880 million phone subscribers. (See fig. 1). In 2002, there were about 45 million; by 2012 925 million. A majority of Indians acquired independent access to information and an ability to communicate that was highly valued and previously unimaginable. Poorer Indians were not unique in valuing the cell phone. In China, researchers concluded that

> the less advantaged ... (people with low levels of education, low income and rural origins) ... felt that the mobile phone widened their communicative and social networks not only quantitatively but qualitatively.[19]

For India, the mobile phone was the most widely shared item of luxury and indulgence the country had ever seen. It quickly became not a luxury but a necessity for tens of millions of people—the single largest category of consumer goods in the country. India's decennial census in 2011 revealed what journalists found remarkable. The country had far more mobile phones than it did toilets of any kind: 53 per cent of the country's 247 million households still defecated in the open; but mobile-phone density in 2012 approached 75 per cent. In hilly states like Himachal Pradesh, where mobile phones saved hours of exhausting travel, teledensity was 82 per cent.[20]

Fortunes, if they were to be made, lay at the bottom of the pyramid (BOP). The wealth that digital technology and Radio Frequency could generate had to come from use by the masses, and such use meant that handsets and mobile connections and services had to be cheap. As one of the prophets of such business plans explained, 'the basic economics of the BOP market are based on small unit packages, low margin per unit, high volume, and high return on capital employed'.[21] India developed the cheapest mobile call rates in the world and turned pre-paid

INTRODUCTION: 'SO UNCANNY AND OUT OF PLACE'

Fig. 1: Phone subscribers in India, 1998–2012, Wireline and Wireless, in millions.

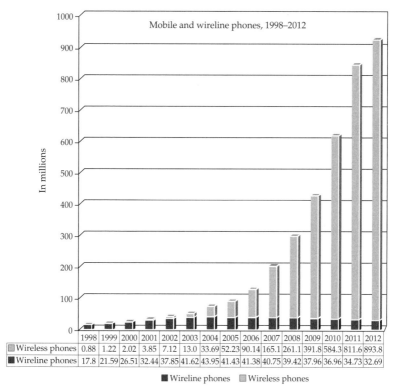

Source: *TRAI Annual Report.*

mobile-phone plans into a complex and much talked-about subject. In 2010, a US dollar (Rs 50) bought more than 200 minutes of talk time on an Indian mobile phone; in Australia, it often bought less than one minute. For the same investment, you drove your phone more than 200 times farther in India than in Australia.[22]

Once the mobile phone reached 'the masses', the masses became consumers. They occupy the third part of our canvas—*Consuming*—in a multitude of ways. Mobile phones were used for business and politics, in households and families and to commit crime and foment terror. Some of the practices enabled by the mobile phone were new and disruptive. At the most fundamental, mobile phones made fishermen's lives safer and sometimes more prosperous. They got warnings of storms and tip-offs about the market for their fish, fewer of which went to waste. Farmers had similar experiences. More information about crops, pests, irrigation and prices was available to more people, and sometimes the 'middle man' could be bypassed and higher prices for crops and livestock obtained (Chapter 5).

But the phone was only a tool. Its effects depended on the knowledge and resources of the people using it, and 'middle men' usually started with advantages that 'lesser' men and women did not share. In politics, the mobile phone was a device that allowed organisations that were already bound together by convictions to exert influence in a manner that hitherto was impossible. The state elections in Uttar Pradesh in 2007, which brought an outright victory to the Bahujan Samaj Party (BSP), owed much to the way mobile phones allowed true-believers to tell the BSP's story and urge sympathetic voters to the polls (Chapter 6). Mobile phones also facilitated crime and terrorism. Indeed, they created new crimes—harassment through text-messaging and faceless frauds where money disappeared without a victim ever seeing the criminal. Mobile phones made vivid pornography more widely available than conceivable in the past. And mobile phones enabled gullible young terrorists to be directed like human drones by remote 'controllers' (Chapter 8).

The ability to broadcast independently—to be an autonomous individual, no longer dependent on literacy, post offices, printing presses or television studios—created a potential to change accepted practices and malpractices, including exposing low-level corruption. Citizens now could record or even video their exchange with an official and use such material to expose slack performance or demands for bribes. Older technology had made video exposés possible. In 2001, for example, the Indian Defence Minister resigned after *Tehelka* magazine carried out a video-recorded 'sting' in which a party functionary was shown receiving a suitcase full of money.[23] But simple technology soon enabled recording to be done even on cheap mobile phones by 'ordinary people'. At such possibilities, bribe-taking officials might tremble or at least have a momentary tremor.

But need they? Through his relationships in Banaras, Assa Doron experienced the highs and lows of the possibilities. In 2009,

INTRODUCTION: 'SO UNCANNY AND OUT OF PLACE'

a local boatman discovered his brother's body hanging in his room. The surviving brother believed it was murder provoked by rivalry on the ghats (riverside mooring areas). With his cell phone camera he took pictures of the scene to convince the police to register a case of murder. (In such circumstances, the police usually aimed to close a case quickly as a suicide.) In spite of his evidence, however, the surviving brother failed. His pictures elicited no sympathy among the police officers he was able to approach, and he lacked the resources and influence to take his case higher up the bureaucratic chain or to the media. A few years later and he would have been able to post his pictures on Facebook or a similar website, and from such exposure greater pressure might have been brought to bear on the police, just as an advertisement for Idea cell-phone services depicted in 2012.[24]

This vignette illustrates both the potential and limitations of mobile autonomy. The boatman had his phone with him—we always carry them with us—when he found his brother's body, which enabled him to capture the scene. But such evidence alone was not enough to force action. Without capable institutions and well-drafted laws, lone citizens are limited in what they can achieve. Shirin Madon illustrated this point with the example of an apparently successful e-governance project that nevertheless was 'dependent on the District Collector—that is, on a special relationship'.[25] Even the well-known online exposure in 2002 of sexual abuse in the Catholic Church in the United States, which Clay Shirky celebrated as an example of the power of the new technology, required the authority of an institution, the *Boston Globe*, to make the campaign of outrage effective.[26]

In the world

One can envisage the mobile phone as a crucial part of a grand theory of social practice and communication. Manuel Castells argued in the 1990s that humanity was experiencing *The Rise of the Network Society* in which 'digital networking technologies ... powered social and organizational networks in ways that allowed their endless expansion and reconfiguration'.[27] The 'material basis' now existed for the 'pervasive expansion throughout the entire social structure' of the 'networking form of social organization'.[28] The cheap mass mobile phone provided that 'material

9

basis', a device by which vast numbers of people acquired the capacity (and the need?) to be part of this 'network society'.

Since the late 1990s, anthropologists have often asked two broad questions about the effects of the mobile phone: 'How do people use this new tool? Does it change patterns of behaviour or simply reinforce existing practices?' To try to get a grip on the variety of global experiences, pioneering research by Horst, Miller, Katz and Ling sometimes drew on concepts labelled 'domestication' and 'performativity' to describe the encounter between humans and technology.[29]

'Domestication' involved the ways in which new devices—in this case, the mobile phone—became part of daily life: how people adapted, appropriated or resisted them to fit with their own needs and circumstances. Many of the essays in the path-breaking *Perpetual Contact* dealt with themes that related to the arrival of the mobile phone in Europe in the 1990s.[30] How and why did people incorporate it into their daily lives? Pure sociability appeared to be one reason—the human desire to interact with others. Other studies treated 'domestication' as a process characterised by ambivalence. As the technology entered people's lives, they had to deal with its varied effects: on household economies, parenting practices, intimate relationships, youth culture and much else. Values and meanings were reshaped in the process: how people regarded 'public' and 'private' or the proper roles of men and women in control of technology.[31]

Curiously, the social aspects appeared to trump the business ones that were originally envisaged as providing the most important market for mobile phones. It is worth recalling that the businessmen who propagated the first telephones in North America took time to discover that the desire for social conversation sold phones more widely and quickly than a wish to do business or to call for help.[32]

Other research into the social aspects of the mobile phone focused on 'performativity'—the way a device became a prop that individuals used in presenting themselves to the world. Everyone has a sense of how the world sees them, and everyone thinks, at least occasionally, about how they are presenting themselves to their world. Unlike a striking haircut, a flashy watch or particular physical mannerisms, the mobile phone not only allowed the individual to make a fashion statement but also to become a broadcast personality and have an audience of thousands.[33]

INTRODUCTION: 'SO UNCANNY AND OUT OF PLACE'

Indians were not alone in their anxieties about the challenges mobile phones set for social practices, religion, government control and state security. Agonised questions were asked everywhere. In Japan, an early refrain was: 'What is this mobile phone doing to our young people?' And in one of the first books about the effects of the cell phone in a specific country: four of the fifteen chapters concerned young people.[34] In India, the question was often: should a woman be allowed to have a mobile phone at all (Chapter 7)? In conservative families, a new bride's mobile might be taken from her on arrival at her husband's household. In its fostering of individual privacy, the mobile phone brought a disruptive intruder into social relationships, gender hierarchies and patriarchal order. In Africa and the Caribbean, on the other hand, a young woman with a mobile phone was a common sight.[35] A character in an African short story says:

> Most of the women who own phones get them from men, who also feed the phones regularly with airtime. ... Sometimes somebody gets into a serious crisis with a wife or girlfriend because he has refused to buy her a phone or to pay for airtime.[36]

In Jamaica, it was widely understood that one of the three common uses of a mobile phone was for 'sexual or potential sexual liaisons'. The other two were for economic transactions and church networking.[37]

Religious use of the cell phone underlines the point that *people* wherever they live invest the mobile phone with meanings that fit existing practices and requirements. People 'appropriate' or 'domesticate' the device, and such appropriations vary across social and cultural settings.[38] The mobile entered the lives of Muslims in Malaysia and Jews in Israel differently, yet in both places religious authorities sought to harness the phone to faith. In Malaysia, anxiety at the corrupting effects of global influences prompted governments to endow the cell phone with a moral character, hence mobile phones had Islamic qualities ascribed to them to incorporate them in everyday Muslim life.[39] To strip the mobile of its Western veneer led to the creation of the first fully Islamic mobile phone, launched by Ilkone in the United Emirates in 2004. The CEO explained that

> consumers nowadays view mobile phones as devices which can add value to their self being and inner feelings rather than just a simple

communication tool. Ilkone i800 is specially designed to serve Muslims all across the world to address their needs, and add value to their spiritual self being.[40]

The phone was welcomed in many Muslim countries. It offered automatic and precise prayer timing, alarms to note the call to prayer, direction-finding to position the user towards Mecca and the full text of the Holy Qur'an with English translations. In Indonesia, mobile-phone service providers in Java offered an array of religious ringtones and special arrangements to cater for extra SMS traffic during religious holidays.[41]

In Israel, orthodox religious leaders similarly battled to shield moral integrity from a mobile-phone market that offered corrupting content, including girls and gambling. The religious establishment used its significant purchasing power to make business heed their demands.[42] In 2005 rabbis joined with one of the smaller mobile operators (Mirs) to launch the 'kosher cellular phone', which religious leaders told their followers to purchase. Supported by a growing number of applications, the kosher phone aimed to supplant the 'sacrilegious' offerings of competing service providers. More than 10,000 numbers for dating services and phone sex were blocked. Calls to kosher phones cost less than 2 cents a minute, compared with 9.5 cents for non-kosher phones, and on the Sabbath every call was penalised with a charge of $2.44 a minute.[43]

Users in India and Sri Lanka also turned the mobile phone towards religious fulfilment. In India screensavers featured saints, gods and goddesses; ringtones buzzed with *bhajans* (devotional songs). Pilgrims carried their cell phones to document their journey, subsequently showing pictures to friends and relatives back home. The album of sacred images and sounds provided evidence of the cultural capital collected on the journey. The digital images enabled those unable to make the journey to gain merit through the sacred vision (*darshan*) of the deity, viewed on the handset.[44] Digitised religion catered for all faiths, including Hindus, Muslims, Sikhs and Christians, and was not restricted to religious material. Mobile phone services also offered images of sexy girls, popular songs and love horoscopes.

In Sri Lanka, the anthropologist Dennis McGilvray discovered the transnational and spiritual significance of the technology when he visited a longtime Sufi Sheikh friend. The Sheikh used cell phones to 'confer blessings and curative incantations on his

INTRODUCTION: 'SO UNCANNY AND OUT OF PLACE'

followers, some of whom live in Denmark and New Zealand'. The scale of his operation was remarkable:

> [The Sheikh] claims he receives 1000 cellphone calls every day from disciples and seekers asking for his divine blessings via cellphone. His followers regularly give him the latest model cellphones as a devout gift to their Sufi master, and he in turn passes his used cellphones on to his favourite followers, who treasure the worn-out equipment that has touched his saintly ears and transmitted his saintly words. He said he has gone through 50 cellphones in the past ten years, each one acquiring *barakat* [charismatic power] from heavy use. Spiritually speaking, this is a win-win form of recycling.[45]

In Indonesia too, 'mobile religiosity', Bart Barendregt concluded, 'has become a business just like any other'.[46]

Not all of the uses for which people deployed the mobile phone were benign. The final chapter of this book looks at ways in which cell phones were used for everything from simple nastiness to unpitying terror. Modern states struggle to monitor and control mobile telephony and prevent users from subverting the state.[47] In China, governments aimed to control the digital flow of information as tightly as possible, yet at the same time to promote a market economy and the adoption of new technology by an increasingly prosperous and educated population. The time-tested method of mutual spying—squeal-on-thy-neighbour—was one way of monitoring. Citizens were invited to email a website or phone a hotline to report illicit trafficking of audiovisual information, including pornography.[48] There were higher-tech methods. 'It is not complicated to add an intrafirm surveillance system over mobile communication—both voice telephony and SMS', wrote a scholar of the Chinese mobile. A company could readily set up a factory-based surveillance system and put a 'wireless leash' on its employees.[49] Governments staged education drives to promote 'traditional values and healthy social norms' among the youth 'as an alternative to the corrupt individual moral practices'.[50]

In conclusion

In the industrialising, urbanising world, individual merit, individual mobility and individual expression have been steadily celebrated for the past 200 years. Pressure of ideas and economics has nudged India in similar directions; but even in the twenty-

13

first century structures of authority—gender, caste, class and family—impose limitations on many people's ability to imagine or conduct themselves independently of such frameworks. This is where the disruptive potential of the mobile phone becomes significant: the new tool affords the possibility of escaping existing structures. People, of course, must already have the imagination to want to do things differently. Only then, as we try to illustrate in the following chapters, do they use the technology to save money independently of a spendthrift husband (Chapter 5), infect others with commitment to a cause (Chapter 6), or communicate with a sweetheart without the approval of the family (Chapter 7). The consequences of such actions are often unanticipated. Their cumulative effects alter expectations about how people should live and present tough puzzles for scholars, policy-makers and politicians.

Mobile phones can both empower and disempower, and it would be a distraction to focus on questions of 'good' or 'bad'. The technology exists; immensely powerful economic forces, augmented by widespread social acceptance, have disseminated it widely; and it will only go away if a major cataclysm befalls humanity. We live with mobile telephony, and most of us relish the benefits. India in this sense is no different from other places. But the disabling inequalities and the diversity of India mean that the disruptive potential of the cell phone is more profound than elsewhere and the possibilities for change more fundamental.

Like Johnny Cash, we want to walk the line, namely that between proclaiming the uniqueness of India on one side and over-emphasising irresistible, universal forces on the other. Easy, private communication introduced tensions and conflicts into households from South Africa to Japan and Finland to Israel.[51] In each place, just as in India, specific histories and ideas shaped anxieties and aspirations. Historically, too, there was nothing new in technological innovation disrupting practices relating to gender and social customs. 'New forms of communication', Marvin wrote of the nineteenth century, 'created unprecedented opportunities not only for courting and infidelity, but for romancing unacceptable persons outside one's own class, and even one's own race'.[52] India's resilient hierarchies have made its encounter with the cell phone more significant, because it gives the lowly poor a potential that has not before been possible for people of their sta-

INTRODUCTION: 'SO UNCANNY AND OUT OF PLACE'

tus. In India, autonomy has often in the past, and still even today, been rationed on the basis of gender, caste and class. Such constraints have been eroding slowly, but affordable mobile communication presents them with greater challenges by offering new and relished opportunities. An Australian resident of Delhi experienced the poignant puzzle:

> On one construction site there were a few labourers, one gentleman looking quite old and having no teeth, probably illiterate, was mixing cement and kept smiling at me. So I went over just to say a few words in Hindi. He had one of the large Samsung smart phones; it was so uncanny and out place. There amongst the dust, pillars and rubble of a building site was this person, dressed very poorly, holding and obviously enjoying his smart phone. He was using one of the applications, but I'm not sure which. Still curious about which application he was enjoying.[53]

Such curiosity—about what is happening, why, how and what it means—fuels our inquiries in this book.

PART ONE

CONTROLLING

1

CONTROLLING COMMUNICATION

I began to sense an unholy alliance among many politicians, bureaucrats and businessmen to stop people from taking power into their own hands through literacy and community-based programs—and through communications.

Sam Pitroda, *Harvard Business Review*, 1993

A little of this book is about subversion and spying, and it is also about governments and high-stakes capitalism. Most of all, it is about ordinary people's lives and how they alter as their access to information changes. At the centre of the story is the mobile phone. To appreciate the remarkable transformation that it wrought in India in the first decade of the twenty-first century, and why this was especially noteworthy, we need to understand how ideas and information were communicated in the past. Only then can we gauge the extent of the change that the mass-ownership of mobile phones made possible. Mass consumption of mobile phones pulverised characteristics of 'old India': the relative isolation of most of its people, the difficulty of movement and the fact that information was more available to the powerful than to the poor and to men more than to women. All these conditions were reinforced by the ideology and practices of caste.

THE GREAT INDIAN PHONE BOOK

Horses, runners and rulers

Control of information and the movement of people have always reinforced power. Before industrialisation began in Britain 250 years ago, most people were peasants, living in villages and subsisting off what they or their neighbours grew. Some people travelled. Alexander the Great led an army 3,000 miles (4,800 km) from Greece to Punjab in the fourth century BCE. Genghis Khan's successors built an empire that stretched from China to Eastern Europe in the thirteenth century. 'The obstacles and dangers of the road', Marc Bloch wrote of medieval Europe, 'in no way prevented travel. But they made each journey an expedition, almost an adventure'.[1] Most people stayed very close to home. If strangers approached your village, the chances were that the consequences would be unpleasant. A few influential people travelled, and they often travelled long distances; but they travelled slowly. Most often, they travelled to make war, to trade or to worship. Today, too, mobiles phones facilitate all three pursuits, but the world of the cheap cell phone is remarkably egalitarian: almost anyone can become part of it.

Language, writing and printing allowed people to create complex tools, organisations and societies. Spoken language is relatively egalitarian: most people learn to speak. Power-holders need language so that underlings can readily follow instructions. Writing is elitist. Even in the twenty-first century, thirty per cent of India—more than 350 million people—could not read or write. In pre-industrial times, literacy was often seen as a skill akin to magic, and rulers of some of India's greatest empires were illiterate, like the Mughal emperor Akbar and the Sikh ruler Ranjit Singh.[2] Akbar, who built the vast empire that preceded British rule, had scribes—reporters—placed in all the provinces. 'Information ... flowed constantly ... to the Mughal imperial centre'.[3] The Italian traveller Manucci described the daily scene:

> It is also a fixed rule of the Moguls that ... the public and secret newswriters of the empire must once a week [send] ... a sort of gazette ... containing the events of most importance. These news-letters are commonly read in the king's presence by women of the *mahal* [harem] at about nine o'clock in the evening, so that by this means he knows what is going on in his kingdom. There are, in addition, spies, who are also obliged to send in reports weekly...The king sits up till midnight, and is unceasingly occupied with the above sorts of business.[4]

CONTROLLING COMMUNICATION

The Mughal empire at its height controlled an area larger than independent India. It did so with communications based on horses and runners, on relatively few literate officials and on a system lacking plentiful paper or the printing press. Information and instructions flowed constantly, but did so slowly. In the 1770s, runners who maintained a good pace carried a message from Delhi to Pune—900 miles (1,400 km)—in 22 days.[5] In the seventeenth century, the travels of the emperor Shah Jahan 'scarcely did more than ten miles a day', and European merchants who used the western port of Surat as their base 'took around ten weeks' to reach Agra, 1,100 kilometres away.[6]

Ordinary people could sometimes travel, often as soldiers or pilgrims. But travel was unusual. A young Brahmin from western India, caught up in the revolt of 1857, had been warned by his father not to leave their village because 'the roads were said to be no good, being full of all sorts of dangers, with deceitful bands of thugs roaming all around'. He ignored his father's advice and did not come home for three years.[7] The great-great-grandmother of R. K. Narayan (1906–2001), on whose legend he modelled *The Grandmother's Tale*, walked from Kumbakonam in Tamil Nadu to Pune in the 1830s in search of a missing husband.[8] When they returned,

> they were carried in two palanquins and had a retinue of bearers who took over in relays at different stages, and many torch-bearers and lancemen to protect them from robbers and wild animals when they crossed jungles in the mountain ghats...[They] arrived about a month later in Bangalore...[9]

Word-of-mouth and gossip were the ways of common communication. 'You may seal the mouth of a furnace', the priest tells Narayan's great-great-grandmother, worried about her daughter's apparent widowhood, 'but you cannot shut the mouth of gossip'.[10]

Rulers recognised that controlling information was essential for maintaining power. They struggled to seal choke-points to prevent valuable news from leaking out from their territories, and they aimed to put the messengers of rivals on their own payrolls so that secret beans might be spilled. A British officer in the state of Awadh in north India in 1844 complained that his instructions from the Governor-General reached the ruler of Awadh before they reached him: 'the identical letter arrived at Lucknow three days before the original reached me'.[11] Being a writer or message-bearer had rewards—payments from various sources—but it also had dangers. When messages and their bearers 'disappeared, it was not always clear ... if bandits ... had killed them or ... a political rival'.[12]

Commercial networks carried information and funds across India with surprising speed. In a celebrated example in 1761, the bankers of Pune were first to know of the Maratha defeat at Panipat 800 miles (1200 km) away—and 'thus managed to limit their losses'.[13] A palanquin, carried in turns by a party of eight bearers, covered between 25 and 40 kilometres (15 to 25 miles) a day, even on poor roads.[14] When Sayyid Ahmed Barelvi, a forerunner of today's *taliban*, embarked on *jihad* from a base in Peshawar in 1830, he had already travelled across north India, collecting funds and followers. After he was killed by the armies of the Sikh kingdom of Ranjit Singh in 1831, two of his followers 'created an effective network for transferring men and supplies, primarily from Bengal [1,400 miles or 2,200 km away], across upper India and to the frontier'.[15] Such things were possible, but they were not common, and they required exceptional circumstances and influence.

Travel and knowledge tended to be confined to political rulers and religious authorities—the people who controlled the sword and interpreted the supernatural. As Bloch wrote of medieval Europe, 'to control a country, there was no other means than to ride through it incessantly in all directions'.[16] In India, social and cultural restraints curtailed the movement of people, particularly poor and low-status people. The caste system of Hindus imposed limits appropriate to people of particular status. Ancient scripture validated the exclusion of low castes from information:

> If a Sudra ... listens in on a vedic recitation, his ears shall be filled with molten tin or lac; if he repeats it, his tongue shall be cut off; if he commits it to memory, his body shall be split asunder.[17]

Low-status people in many parts of India were bound to their places of birth and held in semi-bondage by caste superiors. 'In many cases', writes Sudha Pai, 'they were bonded slaves, often for generations, with no freedom to move out from their villages'.[18]

Brahmins and other high-status people enjoyed wide freedom of movement. Long before British rule or railways, Nambudiri Brahmins from Kerala in the southwest provided priests for the temple at Badrinath in the Himalayas.[19] In Kerala, 'complete freedom of movement', Miller wrote, 'was denied to all but Nambudiri Brahmins, who helped to maintain a unified system of values', which, not surprisingly, validated their pre-eminence.[20] 'Ananda Ranga Pillai never fails to mention in his diary', Deloche writes,

> the brahmins in the service of Dupleix [the 18th-century French adversary of the English] who conveyed the letters from Mahe to Pondicherry [i.e., from Kerala on the west coast to the French centre on the east coast].[21]

Nambudiri Brahmins were 'sometimes used as ambassadors and even for espionage'.[22] In the twenty-first century, mobile phones democratised espionage and subversion, as the skilfully directed mayhem unleashed on Mumbai on 26 November 2008 illustrated (Chapter 8). Scarcely literate youths from rural Pakistani Punjab reported by mobile phones to distant minders as they slaughtered bewildered Mumbaikars.[23]

The East India Company, which slithered into sovereignty over India in the eighteenth century, was founded on ledger books and office procedures. One of its strengths lay in the systematic control of information that a crumbling Mughal empire could not equal. Indeed, when the Company began to incorporate Mughal officers into its apparatus, they were instructed by exasperated officials of the Company 'to number all … letters that it may easily be known if any letter miscarries'.[24] The bureaucratic colonial state quantified and recorded relentlessly. 'The conquest of India', Cohn wrote, 'was a conquest of knowledge'—initially for profit, then for power.[25]

When the printing press began to spread in India from the 1780s, it was scarcely more democratic than the handwriting of court reporters and imperial agents. The East India Company saw the press as a tool of government, enabling the laws and requirements of rulers to be brought to the notice of servants and subjects more quickly than squads of copyists could manage. The press, however, became as big a nuisance for rulers of India as it was for rulers in Europe. James Hicky's *Bengal Gazette*, often deemed the first newspaper in India, was dedicated to attacking the Governor-General, Warren Hastings; it lasted two years before Hicky's press and type were seized by the government while Hicky was in jail for debt.[26]

The printing press, and its offspring the newspaper, were essential ingredients of modern individualism, 'public spheres' and public life. They allowed information—'news'—to be reproduced and to spread widely and quickly. As an English nobleman complained in the seventeenth century, newspapers enabled 'the multitude' to become 'too familiar with the actions and counsels of their superiors'.[27] But printing was costly. 'I calculated', wrote an Englishman setting up a press in Meerut in north India in 1845–6, 'that two thousand five hundred pounds would cover every expense'.[28] This was a huge sum by Indian standards and a substantial one even in Britain at that time. The Englishman and his Meerut-bound press underlined the fact that print and newspapers were not technologies available to everyone. He had to find a press, types, compositors and pressmen and then convey them from Kolkata, first to Allahabad by river barge and then 400 miles (650 kilometres) to Meerut by bullock cart. To produce the paper required improvisation: a tombstone as a makeshift lay-out table and brandy to keep up the spirits of the printers. 'If you were to treat the hands a couple of bottles of brandy, they would stay and set it all up again', the chief printer said after a disastrous first attempt to print the paper. It eventually appeared; but the process was cumbersome and complex.[29] 'Freedom of the press is guaranteed only to those who own one', observed A. J. Liebling. And even then, he might have added, it was not simple and straightforward.

The contest to control the press in India continued throughout British rule. Colonial officials felt contradictory pressures. At one level, they discounted Indian-run publications as unimportant in a country that was 90 per cent illiterate. At another level, however, they worried about the reach and endurance of print and the authority it gave to 'wild rumours' and extravagant criticism. The short-lived Vernacular Press Act of 1878 attempted to control Indian-language newspapers and indicated the anxieties of British governments once printing presses were capable of being easily moved around India on railway trains rather than barges and bullock carts.[30] Governments imposed various forms of censorship on the press during the world wars and the nationalist challenges of the 1920s and 1930s. The most recent example of heavy-handed censorship—as opposed to various other ways of influencing printed material—was the nineteen months of Mrs Gandhi's 'emergency' in 1975–7.

CONTROLLING COMMUNICATION

What has this discussion of the Mughal empire, pre-modern communication, printing presses and censorship to do with mobile phones in the twenty-first century? The answer is simple: it emphasises the contrast—the revolutionary change—between how the mass of people communicated in the past and the way in which the mobile phone transformed the ability to communicate in the first decade of the twenty-first century. Imagine a *Monty Python*-style scene, as the editors of the magazine *India Today* did with a cover theme in 2002. (See Illus. 2). An illiterate Mughal emperor on his mobile phone takes calls from round the provinces, while women of the harem are on their own mobile phones making social calls—no more having to read boring spies' reports to an illiterate ruler. The fantasy catches something significant: the mass mobile phone enabled the illiterate to communicate and the disempowered to connect. The fantasy also hints at the disruption that the mobile phone threatens for relations between men and women. Unless we understand the ways in which information moved in the past, we cannot appreciate the changes in private and public life that the mass mobile phone made possible.

The Government of India built the telegraph from 1851 for reasons of strategy, not profit or social benefit.[31] The telegraph enabled the railways to function safely, and it gave the British government an immense strategic advantage. Rulers needed to control information—to know more than their subjects, just as the Mughals had been well aware 200 years earlier. The great revolt of 1857 was in part 'a struggle ... between the British and the insurgents over the control of modern media of information'.[32] The broad outline of an all-India telegraph system had been completed just before the revolt began, and a last message from British soldiers in Delhi before they were killed in May 1857 alerted British authorities throughout north India to the fact that a major revolt had begun and that the capital of the Mughal empire had expelled the British. An Indian soldier being led to the gallows was said to have gestured towards a telegraph wire: 'There is the accursed string that strangles us'.[33] The mutinying soldiers recognised the threat: 'they ... attacked the telegraph posts and knocked them down', wrote the young Brahmin pilgrim caught up in the revolt.[34] The capacity to exchange information almost instantly represented an immense power, which rulers sought to control tightly.

British governments treated the telephone, like the telegraph, as a device primarily for the state but which others might use for special occasions. (Contrast this with the mobile phone whose most popular use in many cultures is to 'keep in touch' and 'spread the social network').[35] Alexander Graham Bell was said to have made telephony work for the first time in 1876 when he summoned an assistant in a basement at the other end of a wire: 'Mr Watson. Come here. I want you'. When Bell, aged 29, registered the patent, he described the new device as 'an improvement in telegraphy'.[36] Governments in India quickly picked up the technology: the princely state of Travancore in what is today southern Kerala installed a telephone to connect the Maharaja's palace with the government offices in 1878.[37] The Government of India licensed private companies to provide telephone services in Chennai, Mumbai and Kolkata in 1881. Mumbai and Kolkata had exchanges operating a year later. The Kolkata exchange in 1882 had 50 lines, connecting officials to officials.[38] The telephone was a tool of government, nicely illustrated by the first automatic exchange, not in one of the great cities, but in the summer capital, the hill station of Shimla.[39]

For a foreign government, the telephone seemed a controllable medium. It did not send messages to multitudes; it needed expensive equipment and government-maintained wires to operate; and anyone who owned a phone had to be known to the government. It lacked the capacity to reach large numbers of people, whether to arouse them or sell to them. These qualities may have made the phone more palatable to M. K. Gandhi. The 'father of the nation' did not like radio or the movies and was suspicious of the advertising that dominated most newspapers. The telephone, however, aroused no such fears. In his *Collected Works*, there are frequent references to his talking on the phone, and he was happy to be photographed with a phone in hand. He gave explicit and thoughtful instructions about how the telephone box at the Sevagram ashram was to be designed.[40] He showed remarkable comfort with the telephone for a man who had written in *Hind Swaraj*:

> Machinery is like a snake-hole which may contain from one to a hundred snakes. Where there is machinery, there are large cities; where there are large cities, there are tram-cars and railways. And there only does one see electric light. Honest physicians will tell you that where means of artificial locomotion have increased, the health of the people

has suffered. I remember that, when in a European town there was scarcity of money, the receipts of the tramway company, of the lawyers and of the doctors went down, and the people were less unhealthy. I cannot recall a single good point in connection with machinery.[41]

The Gandhian attitude may explain two aspects of Indian approaches to telephony. First, it was considered benign and endorsed by Gandhi as a way an inveterate letter-writer like him might communicate person-to-person and get instant replies. Second, however, it was to be treated as a rare device used only for important purposes. He stipulated in designing the phone box for the ashram that

> the telephone cabin should be located outside the office, so that the telephone would be accessible even when the office is closed and ... the telephone used whenever necessary.[42]

Such an attitude helps to explain why the telephone was treated like a beneficial colonial legacy after independence, something that helped government and business and might in emergencies be availed of by the general population.

When India awoke to life and freedom in 1947, it had about 100,000 telephones for a population of 340 million people: one telephone for every 3,400 people.[43] Growth was not a priority, and it came slowly. Bombay, India's commercial centre, had an 'outstanding demand for telephones' of 34,000 in 1955 when the Communications Minister announced what the *Times of India* described as 'Big Plan for "Phones in Bombay City"'.[44] Five years later, however, 'the telephone service goes on deteriorating' lamented the *Economic Weekly*, and 'the waiting list of applicants ... goes on lengthening ... If this is the state ... in India's premier city [Bombay], one can imagine the condition in the rest of the country ...'[45] By 1970, the national waiting list for a telephone was estimated at 700,000 with a waiting period of more than four years. 'The prospects of easy availability of connections even ten years hence appear bleak'.[46] By 1987, the waiting list stood, with mystifying precision, at 842,567; and 3.3 million of the installed phones were deemed to be 'outdated instruments'.[47]

The spread of telephones marched a little ahead of the growth of the Indian economy for the first forty-four years of independence prior to the 'liberalisation' that began in 1991. By that year, India had 5.1 million phones, an increase of just under 10 per cent a year from 1948.[48] And the number of people queuing around each phone to make a call had improved—there was now one phone for every 165 people. (See Table 1.1). But many of these phones, as we know, were in government offices or businesses, often in locked boxes to be used only for official purposes after a would-be caller had been provided with the key.

Table 1.1: Phone Connections in India, 1947 to 2011.

Year	Number of phones in '000s	People per phone
1947	100	3400
1964	580	800
1984	2,600	280
1991	5,100	165
2001	37,000	28
2011	900,000	1

Sources: *Statistical Outline of India* (Mumbai: Tata Services) and *India: a Reference Annual* (New Delhi: Publications Division) for relevant years.

How did 'ordinary people'—the overworked *aam aadmii* of Indian political rhetoric—communicate? Before the British, as we have seen, they communicated by word of mouth, transmitted by officials, merchants, soldiers, pilgrims and marriage parties. The colonial regime brought industrial technology: the printing press and then, more significant for ordinary people, the post office, the telegraph and the railway. These three, all government run, became the greatest employers in the country. When Jawaharlal Nehru died in 1964, seventeen years after independence, Indian Post and Telegraphs employed more than 400,000 people, ran close to 90,000 post offices and carried more than 4.8 billion items of mail a year, roughly ten pieces of mail for every person. Indian Railways employed 1.2 million people and carried 500,000 passengers a day. Telephones, however, barely found a place in these statistics of communication. In 1964, there were about 580,000 phones for a population of 465 million people—roughly one phone for every 800 Indians.[49]

Communication was slow and patchy in these post-independence years. Newspaper circulations began to grow strongly only from the 1980s. Until 1979, daily newspapers in English sold more copies than those in any Indian language, including Hindi, the official language used by more than 40 per cent of the population. Radio was still closely controlled by government. In the early 1960s, 20 per cent of programming consisted of classical vocal and instrumental music. Radio required electricity. Battery radios were large and expensive until the arrival of the transistor. When Nehru died, a population of 440 million had 3.6 million radios—one for every 120 people.[50] Television did not become an all-India medium until the Asian Games of 1982.[51]

Even in 1991, 57 per cent of India could not read or write. Most people communicated through the spoken word sometimes mediated by people reading aloud or by listening to cassette tapes which had become available for entertainment and for political purposes from the 1970s.[52] The postcard was the closest the majority of people came to communicating over distance. The basic postcard cost 6 pies in 1953 (one-32nd of a rupee or half a US cent) and rose only slightly to 50 paise (half a rupee or one US cent) in the twenty-first century.[53] Postcards conveyed essential news, written, if one were illiterate, by a professional writer or a literate neighbour, as Shyam Benegal's popular Hindi film *Welcome to Sajjanpur* (2008) wryly captured.[54]

The postcard was a wonderful device. It allowed cheap communication, but for illiterate or semi-literate people it required intermediaries, and it was slow, uncertain and open to prying eyes and sabotage. Its heyday was probably in the 1960s and 1970s when the volume of posted items grew strongly. The post office reached a peak when it carried 7.4 billion items annually at the beginning of the twenty-first century, but as the mobile phone spread, postal volumes fell—by more than 10 per cent between 2004–05 and 2008–09.[55]

Table 1.2: Items carried by Indian Posts, 2004–05 to 2008–09.

Year	No. of items carried '000,000s
2004–05	7,360
2005–06	6,701
2006–07	6,677
2007–08	6,391
2008–09	6,541

Source: Department of Posts, *Annual Report* (New Delhi: Department of Posts, India, for relevant years).

The national elites, who ran India after independence, proclaimed Gandhian values. The telephone, like television and even radio, was regarded either as a luxury or as an item for the strategic use of the state, similar to the *dak* (postal system) of the Mughal empire. An advertisement for Bharat Electronics (BEL) in 1964 reflected this attitude. Headlined 'The Message That Saved His Life', it drew on the patriotism and fears of the disastrous 1962 border war with China and showed a soldier making a call on a enormous radio-telephone while colleagues carried a stretcher towards a helicopter. The copy read:

> He was wounded seriously in one of the forward posts. The nearest hospital was 30 miles away. A message was flashed by means of a BEL-made high frequency transceiver ... The life of yet another of the nation's brave sons was saved! ... BEL is geared up to meet the nation's needs.[56]

The wounded soldier of the advertisement was lucky: had he been injured in a village accident, thirty miles would have been the likely distance to the nearest telephone, much less the nearest hospital.

The elite nature of the Indian state after independence was not surprising. Caste and colonialism were complementary. Though independent India happily embraced elections, law codes, egalitarian rhetoric and the anonymous intimacies of railway travel, the country remained highly stratified, and holders of political power came from a tiny segment of society.[57] The practice of untouchability was banned in Article 17 of the constitution of 1950: '"Untouchability" is abolished and its practice in any form is forbidden'. But Article 17 collided with the ancient texts that promised molten metal in the ear and the cutting out of tongues for low-status people who listened to or passed on high-status knowledge.[58] Long-standing social practice and prejudice wrestled with 'modern' law after 1947. Affirmative action—'reservation' as it was dubbed in India—provided seats in legislatures for Scheduled Castes ('untouchables') and Scheduled Tribes, as well as places in schools and colleges and in government jobs. Similar reservation of government jobs was created for Other Backward Classes (OBCs). More than sixty per cent of the population of India was made up of people who did not belong to higher castes.

The people who drafted the constitution of 1950 and ran the national government were largely upper-caste lawyers who knew

English. By 2012, India had had only two Prime Ministers from lower castes, the brief tenures of Charan Singh in 1979 and H. D. Deve Gowda in 1996. The sense of 'knowing what was good for the masses' stemmed both from the civil-service mentality inherited from the British and from the notion that higher castes were the arbiters of what constituted suitable conduct for society. The development of radio provided an example. The British treated radio with suspicion and controlled it tightly. After independence, high-caste government ministers made All India Radio (AIR) a byword for ponderous high culture, government pronouncements and tedium.[59] Maintained as a department of government, radio and television were managed in ways that discouraged innovation and ensured that ordinary people only rarely found programming that connected with their lives.

The First Five-Year Plan in 1951 recognised the need for better telephone service:

> The most insistent demand for development is in the fields of telephones. India is very backward in this regard; it is even more backward than, for instance, China. The number of telephone lines in the whole of India is less than that in one city of Australia, vis., Sidney [sic].[60]

But grand economic goals of self-sufficiency and elimination of imports meant that telephones waited for government to attend to them. 'For the rapid development of the telephone service', noted the Second Five-Year Plan in 1956, 'it is necessary that telephone equipment should be manufactured within the country'. A factory was built that could produce 50,000 phones a year by the mid-1950s. It started 'by assembling telephone instruments from imported parts', but later was able 'to produce 520 out of the 539 parts of a telephone instrument'. The discussion continued in a manner characteristic of the time. The product and what it might do were not so important as the fact that it was 'made in India'. 'Of the remaining 19 items as many as 17 are manufactured by other Indian firms, only 2 being imported from abroad'.[61] The Five-Year Plans, concluded one analyst, 'gave telephones low priority, classifying them as a luxury'.[62] Such an attitude was not unique to India. Elites elsewhere tut-tutted at the acquisition of mobile phones by poor people. 'Wealthier Jamaicans', Horst and Miller wrote, 'often denigrate impoverished individuals for continuing to buy and use cell phones'.[63]

The big black bakelite blocks that were telephones in pre-mobile India could be instruments of exquisite torture. They were carefully guarded, often in locked boxes. Lines were often out of order, and if they were working, they crackled like pine needles in the open fire of a hill-station guesthouse. To locate the appropriate person in an office in which one phone served dozens of people called for the patience of Job and the luck of the Irish. Placing a trunk call could involve half a day of waiting for the operator to ring back, frequently with the message that the recipient was unavailable or that the line was out of order. For extra payment one could place a 'lightning call'. 'Using a telephone', reflected one Indian writer, 'used to be a hellish undertaking'.[64]

All this is not surprising given the attitude to telecommunications of some of the elite—who did have access to phones. A document intended to initiate discussion about the Sixth Five-Year Plan in 1977 argued:

> The primary need of the people is food, water and shelter. Telephone development can wait. In place of doing any good, development in the telecommunications infrastructure has tended to intensify the migration of population from rural to urban areas. There is need to curb growth of telecommunication infrastructure, particularly in the urban area.[65]

In 1977–79 and 1989–91, coalition governments imbued with belief in small-scale self-sufficiency reduced expenditure on telecommunications in the national economic plan.[66] When Mrs Gandhi was murdered in 1984, there were fewer than four telephones for every 1,000 people: 2.6 million phones for a population close to 730 million. (See Table 1.1).[67]

Untying communication

A modest change began when her son, Rajiv Gandhi, became Prime Minister. Journalists gave him a reputation for being a lover of gadgets, but he appears simply to have been up to date. He 'introduced a new style of work at the highest level'. Powerpoint presentations replaced typed memoranda. This was 'absolutely novel in 1985–86. Everyone was free to present his point of view but only through transparencies using an overhead projector or a PC and LCD projector'.[68] As a former airline pilot, Rajiv Gandhi had regularly faced the consequences of a paralytic telephone system.[69] Airports and airlines depended on telephones for activities ranging from bookings and passenger alerts to organising crew and finding hotels. Within two months of becoming Prime Minister, his government created a new Department of Telecommunications, hived off from its old association with the Department of Communications and the post office.[70] By 1986, two public corporations had been created with the intention that they would work like businesses rather than government departments. One was to handle overseas communications (VSNL—Videsh Sanchar Nigam Ltd), and the other was dedicated to the telecom needs of the two great cities, New Delhi and Mumbai (MTNL—Mahanagar Telecom Nigam Ltd).

The noticeable expansion of the ability to share information—the democratisation of information-sharing—came with the creation of tens of thousands of Public Call Offices, affectionately known as 'PCOs'. (See Illus. 3). Public telephones of any kind had been rare. Post-office phones intended for public use were often out of order, and in the queues waiting to place calls people had time to play cards, write letters and form life-long friendships. Sam Pitroda, a US-based Indian engineer and entrepreneur, was sometimes credited with having made the PCO concept possible, both technically and bureaucratically.[71] 'I was captivated by the man', wrote one of the admirers who worked with him, 'and life was never the same again for me'.[72] The rags-to-riches version of Pitroda's life tells of a man who made a fortune in the US and then determined to do something for India. He convinced Prime Minister Indira Gandhi of his vision and became a friend of Rajiv Gandhi. Indira Gandhi enabled Pitroda to set up the Centre for Development of Telematics (C-DOT), and during Rajiv's time in office from 1984–89 Pitroda and C-DOT flourished. The achievement lay in designing, and getting into production, small cheap digital switching units—the boxes that transfer telephone calls from the caller to the recipient. 'These rural exchanges', Pitroda wrote proudly, 'were small masterpieces of "appropriate" design', able to withstand dust, heat and rough handling.[73] Their manufacture was licensed to private companies, one of which later transformed itself into Airtel, India's biggest provider of mobile-phone services in the early twenty-first century.

C-DOT could claim only tangential responsibility for mobile-phone expansion, and C-DOT's detractors pilloried Pitroda after

Rajiv Gandhi's defeat.[74] Mobile phones took another fifteen years to trickle, and then flood, across India. C-DOT could, however, claim an educational achievement. Its 'switching units' drove the creation of Public Call Offices (PCOs).[75] Pitroda wrote that 'in 1964 … I was only 21 years old and I had never used a telephone'.[76] PCOs not only lessened the horrors of making a phone call; they put phones within reach of people who had never used one before. In 1980, there were only 12,000 public phone booths in the whole of India. Once it became possible through electronic rather than mechanical switching of calls to handle larger volumes of traffic, PCOs were encouraged as small businesses. Painted bright yellow, the booths sprang up around the country from 1986 either as part of existing shops or as new enterprises; the right to run a PCO often was granted to people like ex-servicemen. A simple meter recorded the length of a call and calculated the charge. PCOs did not penetrate the countryside; there was not enough traffic in villages to make the business worthwhile; but now it was possible to place quick reliable calls to phones throughout India. By 1990, a student living away from home, as Pitroda was in 1964, was no longer likely to be a stranger to the telephone.

In raw numbers, however, the telephone was still a rare medium in India at the time of the First Gulf War and Rajiv Gandhi's assassination in 1991. The population of 840 million people had 5.1 million telephones—the imaginary queue to use a phone was down to 165 people for every telephone (Table 1.1). Though PCOs meant that it was possible to place calls reliably and with perhaps only a few minutes of queuing, the telephone was still a special instrument, its ownership and daily private use confined to urban upper-middle classes, businesses and officials. In 1994, the waiting list to get a telephone installed was four years, and the number of people on it had grown to two million.[77]

Questions of chickens and eggs arise when one analyses the growth of Indian telephony in the 1990s. It is an old and much-loved debate among the social scientists. Does technology mould behaviour or do people and their circumstances determine why and how technologies are accepted or rejected? Invention is essential. But the existence of a device does not mean large numbers of people will adopt it. A cluster of things happen to make technologies enter people's lives in widespread life-changing ways. This happened, as things often do in India, slowly.

CONTROLLING COMMUNICATION

The small changes of the 1980s represented the recognition by Rajiv Gandhi and his government that telephones needed to work better. Internationally, the United Nations declared 1979 a year of telecommunications, and advocates of development increasingly emphasised improved communications as crucial for bringing benefits to the poor. All over the world, government-controlled monopoly telecommunications organisations were being broken up.[78] As India's overseas population grew, prospered and moved back and forth to India from the US, Canada and the UK, they wailed at the poor quality of the telephone system.

Two major obstacles needed to be overcome: the need for wires and the inertia of the state-controlled telephone monopoly. With 75 per cent of the population in 600,000 villages, the difficulties of connecting large numbers of people—or even large numbers of villages—to a telephone line were immense. A 1970 estimate calculated that 8,500 kilometres of cable were necessary for every 100,000 telephones installed. India produced less than half that amount of cable annually, enough to install perhaps 50,000 phones. Having designated the manufacture of telephone cable as something to be done by public-sector industries, the government decreed that cable would not be imported.[79]

Various innovations in the technology and governance of telecommunications were in progress elsewhere in the world when India's heavily regulated economy was hit by the freight train of the First Gulf War. The effects of the war knocked the old economic system off the rails. Inflated oil prices and devastated foreign-exchange remittances from Indians working in West Asia threatened to leave India too poor to pay for its energy imports. Elections came in the midst of this economic crisis, and during the elections Rajiv Gandhi was assassinated. A Congress-led government came to power in June 1991 and quickly agreed to dismantle a number of economic controls in return for loans and guarantees from the World Bank and the International Monetary Fund. Under this pressure, Indian governments began to loosen the bureaucratic straitjacket that bound the economy. Telecommunications was one of these areas. Indians in India and overseas desperately wanted a better phone system; international capitalism wanted a freer Indian economy.

The preamble to the National Telecom Policy of 1994 (NTP-94) captured the themes:

> The new economic policy ... aims at improving India's competitiveness in the global market and rapid growth of exports[,] ... attracting foreign direct investment and stimulating domestic investment ... It is, therefore, necessary to give the highest priority to the development of telecom services ...

The objectives of the new policy were 'telecommunications for all, ... universal service covering all villages, ... telecom services ... of world standard' and the transformation of India into a 'major manufacturing base and major exporter of telecom equipment'.[80] NTP-94 noted the scale of the task. It estimated that there were 8 phones for every 1000 people. The waiting list for a phone had grown to 2.5 million. Only one in four of the 580,000 villages had a telephone (and was it working?). And in spite of the improved accessibility resulting from the Public Call Offices of the late 1980s, there were still only 100,000 PCOs for a country of 900 million people. The new policy aimed to eliminate the waiting list, put phones in every village, have 1.9 million Public Call Offices by 1997 and 'raise the telecom service in India to international standard'. How was this to be done? 'Private investment ... would be needed in a big way to bridge the resource gap'.[81]

As well as basic landline telephones, the new policy listed six services that private companies had in theory been allowed to provide since 1992. The last of these—it sits in the policy document almost as an afterthought—proved to be the addictive one: 'cellular mobile telephone' service. To provide such a service, government had to allot scarce Radio Frequency spectrum to private companies, and this was to be done 'through a system of tendering'.[82]

US governments had faced a similar problem: how, and to whom, was this invisible, immensely valuable and seemingly magical commodity, Radio Frequency spectrum, to be allotted? In the US the Federal Communications Commission (FCC) auctioned the right to radio frequency in hundreds of small geographical parcels in the 1980s. The result at one point was 92,000 bids for 180 geographical areas. The FCC threw up its hands and awarded rights by lottery: lucky winners, often small-time players, sold their rights at a profit to bigger organisations.[83] Indian governments in 1994 similarly proposed to call for bids, and a great corporate game began, punctuated by intense bureaucratic politics and allegations of corruption. Senior bureaucrats, crafty

CONTROLLING COMMUNICATION

politicians and capitalists large and small wrestled to control this invisible asset.

The Mughal emperor kept tabs on his subjects through a system of employees and reporting. Every town in seventeenth-century India had a *kotwal*, a variation of a magistrate or police chief. The *kotwal* employed the town's scavengers who cleaned household latrines. These people, wrote the Italian traveller Niccolao Manucci, 'go twice a day to clean out every house, and they tell the *kotwal* all that goes on. On his part, the *kotwal* must render an account to the king of what he has heard has happened, whether it be by night or by day'.[84] Information was gathered piece by piece—'sweeper's wisdom', a modern magazine editor called it.[85] Each scavenger knew the comings and goings for the houses that he or she covered; but only the *kotwal* sat at the centre of the town's web of information. And the emperor's officials sat at the centre of the imperial web stretching from today's Afghanistan to Bangladesh in the east and as far south as the River Kaveri.

What would have happened if every scavenger had been able to communicate with every other scavenger? The capacity of 'lower orders' to know about the doings of their superiors would have been greatly enhanced. Freer movement of people and information contributes to egalitarian ideas and practices. All-weather roads did this in Europe and elsewhere from the eighteenth century. The cheap mobile phone, capable of being owned by hundreds of millions of individuals, was an even greater disrupter than the King's all-weather highway.

No tool is neutral. Its purpose and politics come from the people who use it. All-weather roads could be used to help young Scots go to London and seek their fortunes, as Samuel Johnson said. But all-weather roads could also bring taxation, conscription and the armies of 'Butcher' Cumberland to peasants' doors.[86] Similarly, the benefits of mobile telephony vary with wealth and power. The processes that the NTP-94 set in train launched India's greatest capitalists and most powerful politicians and civil servants into competition to control Radio Frequency spectrum for the immense wealth it could bring.

2

CELLING INDIA

Make a phone call cheaper than a postcard and you will usher in a revolutionary transformation in the lives of millions of Indians.

Dhirubhai Ambani, in Keskar, 'Reliance Infocomm'

A huge Indian flag flutters over the memorial where Rajiv Gandhi was murdered by a suicide bomber. The village of Sriperumbudur is about 30 kilometres west of Chennai, and Rajiv Gandhi's electioneering cavalcade reached the place about nine o'clock on the evening of 21 May 1991. He had been riding in his white Ambassador car—a symbol of out-dated self-sufficiency in post-independence India—with Barbara Crossette, the correspondent of the *New York Times*. Moments after he left the car there was 'a large boom' and he was blown to pieces. Crossette was hustled away in the same car, back towards Chennai. 'All along the route it was clear that no one had heard of the assassination. It took the news agencies nearly an hour to begin telling the people of India that Gandhi was dead'.[1] In New Delhi another reporter was celebrating a birthday with other journalists when a message came about 11pm that the former Prime Minister had been assassinated. 'I went upstairs', he wrote, 'and took one look at the Press Trust of India and United News of India teleprinters and they were carrying one-sentence flashes ... All India Radio didn't carry the story, I don't think, until midnight'.[2]

In 2010, the largest cell-phone factory in India and one of the largest in the world stood two kilometres from the Rajiv Gandhi memorial. The Nokia factory at Sriperumbudur was located in an 85-hectare Special Economic Zone, where 12,000 workers produced thousands of phones a day. The Ambassador cars, teleprinters and peaceful obliviousness that Crossette saw in the countryside around Sriperumbudur in 1991 had been replaced by a Hyundai car factory, Korean restaurants, pervasive mobile phones and steady streams of buses ferrying workers back and forth to the Special Economic Zone where various industries, as well as Nokia and Hyundai, had been set up.

When Rajiv Gandhi died, there were six phones in India for every 1,000 people, and Crossette had to wait till she got to a hotel phone in Chennai to tell her newspaper what had happened. Late at night on the road from Sriperumbudur in 1991, it made no sense to look for a working phone. A car-based mobile phone had been introduced in New Delhi in 1986, but by June 1988, 'the number of subscribers working in Delhi are 76 and there [was] a waiting of 252'.[3]

If the coming of the mobile phone to India were told as a fairy tale, it would begin with a description of a Wicked Witch holding her subjects in bondage by preventing them from talking to each other. Indeed, such a picture is not too different from the one Ashok Desai presents in his path-breaking book *India's Telecommunications Industry*. When Rajiv Gandhi died and India's economic reforms began in earnest, the telecommunications industry was 'a placid backwater' where 'the service was poor' and a telephone connection had as one of its chief virtues a 'black-market value'. A phone was a luxury that very few people had, and, if they did, a phone's use for actually making calls was limited and uncertain.[4] In the years that followed, the Wicked Witch, in the guise of the telecom bureaucracy of the Government of India, used all her wiles to retain control of the invisible charm called Radio Frequency spectrum, which was acquiring great value. In the end, the Witch had to surrender some of her powers and the people of India found themselves able to communicate as never before.

This caricature captures aspects of the struggles over the use and allocation of Radio Frequency after 1991. That drama had four acts. It began in 1991 with attempts to deregulate the Indian economy and allow private industry to provide telecommunications alongside the government provider. This episode culminated in the tortured labour and birth of the Telecom Regulatory Authority of India (TRAI) (intended to be the referee for the new competitive telecom industry) in 1997. The second period could be called 'The Travails of TRAI' as the new-born regulator struggled haplessly with the telecommunications bureaucracy of the Government of India. The bureaucracy—multi-layered and maddeningly patient—acted with remarkable coherence to hobble the regulator. This standoff led the Prime Minister, Atal Bihari Vajpayee, to take the telecommunications portfolio into his own office in 1999 and abandon many of the failed rules and policies of the previous five years. A New Telecom Policy was announced in 1999 and over the next two years most controls on private companies providing telecom services were removed.

Until 2002, expansion of the telephone network across the country had been modest. There were only 45 million phone connections for 1,000 million people. By 2012 telephone connections reached 900 million.[5] The robber barony and institutional skulduggery that went with this expansion form the third part of the story. Act IV, in which ministers were jailed and the courts abrogated agreements that had attracted huge investments, played out well into the twenty-first century.

Act I: '… Within a fortnight …'

The above assurance—'within a fortnight'—given to would-be telephone subscribers in the early 1990s carried a melancholy promise of long waits and musty files. The same document spelled out how such marvels would occur:

> applications for new connections received along with necessary deposits will be registered in the office and entered in a register on the very date of receipt and entries attested by the officer authorised for the purpose.
>
> … A serial number will be allotted against each application. Serial numbers in the Waiting List should be given in the order of the date of payment. … The registration number shall be intimated to the applicant within a fortnight in the pro forma attached to the application form.[6]

Even the promise that applications would be registered 'on the very date of receipt' was little comfort when the waiting list ran to 2.5 million. *Swamy's Treatise on Telephone Rules*, the 850-page guide to the procedural mysteries of the government telephone monopoly, was a profitable publishing venture as late as 1993 (See Illus. 4).

Why had the telephone spread so little in the forty years after independence? There were physical reasons. The country was vast and rural. To connect 600,000 villages by copper-wire phone lines would have been a huge and expensive task. Copper was valuable. A meter of cable contained more than three kilograms of copper, and copper in the 1970s was worth about US $4.00 (Rs 30) a kilogram. A metre of copper phone cable in those days could add up to more than Rs 100: a week's wages for a school teacher or lower-ranking government servant.[7] Not surprisingly, rural phone wires sometimes disappeared.

In 1990–1, the First Gulf War's jolt to the economy increased oil prices which drained foreign exchange and cut off remittances from workers in the countries of West Asia. The war threw tens of thousands of Indians out of work in the Gulf, many returned to India, and the foreign exchange they had been sending home dried up. India was running out of dollars to pay its energy bills.

Telecommunications in India, and particularly telephones, were ripe for change. The exasperation of a growing middle class coincided with global economic and administrative pressures to reduce government controls on the economy. Such pressures, combined with the opportunities offered by new technologies, put telecommunications near the top of the reform list of the Congress-led government that came to power in June 1991. The Public Call Offices had allowed large numbers of Indians to experience the convenience of a functioning telephone for the first time. Such experience increased the frustration at the difficulty in obtaining a household or office phone and at the unreliability of a phone once it was installed.[8]

A strike by government telecom employees in December 1990 emphasised the need for change, even to sceptics. The corporatised government entity that provided telephone services to New Delhi and Mumbai (Mahanagar Telecom Nigam Ltd or MTNL) proposed to pay its 70,000 employees a bonus for good performance. MTNL had been hived off from the Department of Communications in 1986 as one of Rajiv Gandhi's attempts to develop more efficient telecom services. The proposed bonus was intended to reward workers in the new entity for steady work in meeting targets. But the 370,000 employees of the *other* government telecom agencies went on strike for a similar payment, though no one except they themselves was suggesting they had worked notably hard or well in the previous twelve months. In the end, no one got a bonus, but the caretaker government, which ran the country prior to the 1991 elections, set up a Telecom Restructuring Committee under M. B. Athreya, a Harvard-trained management specialist.[9] Thus when the new government under P. V. Narasimha Rao and with Manmohan Singh as Finance Minister came to power in June 1991, it had the report of a review committee waiting for it. The Athreya Report floated easily on the tide carrying the government towards economic deregulation.

The recommendations of the Athreya Committee were never going to be popular with telecom bureaucrats or 400,000 government employees of the monopoly telecom services. It recommended the creation of a new entity to provide services, in the way that MTNL was already doing for the metropolises of Delhi and Mumbai. Such a new corporation would be government-owned but independent of the Ministry of Communications.[10] It recommended allowing private enterprise to provide a limited range of telecom services, including what was then thought to be a luxurious frippery for India, mobile phones. To oversee these changes and act as a referee, the committee called for the establishment of an independent Telecom Regulatory Authority.[11] All three recommendations were eventually implemented; but it took ten years and a tortured series of contests in which a government department—the Ministry of Communications—often appeared to be resisting the policies of the government it was supposed to serve.

In the midst of its financial crisis and the need for World Bank and IMF support, the Narasimha Rao government began to open the economy to private interests and even foreign investment. It called for bids from private companies to provide mobile phone services—but not landline services—in the four big cities, Mumbai, Delhi, Kolkata and Chennai (or Bombay, Delhi, Calcutta and Madras, as they then were). Two companies were to be licensed in each city; they were required to have a foreign minority partner with telecom experience; and bids were to be evaluated largely on the basis of the share of revenue the company undertook to pass on to the government.¹² According to one account, the telecom ministry's officials, fighting a public-sector rearguard, 'regarded mobile telephone service as a value-added' that would simply route more business through their monopoly landlines.¹³ In any case, this first attempt to bring competition into telephony got bogged down in the courts when unsuccessful bidders challenged the bidding process; licences were issued for these limited big-city services only at the end of 1994.¹⁴

In May 1994, when India still had fewer than 9 million phones and the courts were puzzling over challenges resulting from the first changes in telecom policy, the government issued a National Telecom Policy (NTP-94). Critics claimed it 'clearly reflected the dominance' of the government servants in the telecom ministry.¹⁵ The National Telecom Policy of 1994 emphasised the need to provide phones on demand (not after a two-year waiting period) and to put a phone in every village. But it pleaded that the cost 'clearly … is beyond the capacity of Government funding'. Thus, almost regretfully, 'the private sector would be needed in a big way'.¹⁶ NTP-94 allowed Indian businesses, in partnership with experienced foreign telecom companies, to bid for the right to provide services in twenty-one geographical circles largely corresponding to India's states.¹⁷ (See Map 2, p. xxxiv). The new entrants were to be allowed to offer 'basic telephone services'—landlines and other services previously provided solely by the government telephone department.

For those involved with the Athreya Committee, this was lukewarm reform. It did not break up the various functions of the Ministry of Communications. As things stood, the ministry determined policy (it laid down the rules), acted as the arbiter over policy (it interpreted and enforced the rules) and ran the telephone system (it installed telephones in homes and businesses). The Athreya Committee advocated creation of both an independent regulator and a separate corporation to take over what had been the government monopoly on the installation of phones. Such an entity, the unions feared, might be ripe to be sold off to private shareholders. The Athreya Committee also called for the opening of all telecom services to private enterprise and competition. Over the next ten years, much of this came to pass, but changes were, in Athreya's words, 'opposed, diluted and emasculated from within the Department of Telecommunications'.¹⁸

Once the Indian government decided in 1994 to open some of telephone services to private-enterprise providers, it had to decide on what basis to allocate Radio Frequency spectrum—how to choose the lucky winners. Elsewhere in the world, two methods had been followed: auctions and 'beauty contests'. In an auction, governments make known the amount of spectrum on offer and the requirements that successful bidders must fulfil. In theory, companies making the highest bids are awarded use of the spectrum for a fixed period. In a 'beauty contest', aspirants submit proposals explaining their capacity to meet various criteria of service and payment, and an 'expert committee' selects the winners.[19] In 1994, the government of India, and its Communications Minister, Sukh Ram (b. 1924), decided to auction licences to private companies that would be allowed to provide some, but not all, forms of telephone service.

'The auction process', one analyst concluded, 'was a mess'.[20] Critics sympathetic to the old central-planning regime branded the government as cash-strapped, beholden to foreign agencies and 'mainly interested in raising revenue by auctioning the licence rather than in expanding services'.[21] They predicted—gloriously wrongly—'in a country where people do not regard the telephone as a necessity, the expansion of the network will slow down considerably with higher tariffs', which were likely to result from private providers having to pay government large licence fees derived from a small base of users.[22] They were right about fees: high fees limited subscriber enthusiasm; but they were wrong about people's desire to have a phone, if the price was right.

The messiness of the auction process, which began in 1995, had two dimensions. The first arose from the very high amounts that bidders pledged for licences. The Department of Communications, which was still rule-maker, rule-interpreter *and* provider of phones, put stringent conditions on what a private operator could do with the spectrum it was allocated—and what it had to pay to connect its clients to the government-controlled telephone system. The government system was still the biggest; it still owned nearly all the landlines; and it was still the only system to cover, in its own dilapidated way, the entire country.

The conditions imposed on new private operators had the 'effect of ensuring their unviability'.[23] They were permitted to charge their subscribers no more than Rs 156 a month, yet to meet their licence payments to government and build infrastructure for an effective network, they would have had to charge probably eight times as much. Moreover, if customers in the government network called a telephone in the private-operator network, the receiving party, not the caller, was charged.[24] To try to cover the gap, the new companies charged the highest permissible rate for each call a subscriber made—a deterrent to users and new customers. 'The conditions of their licence ensured that their revenue would fall short of their costs',[25] Desai asserted. People who used a cell phone in the 1990s recalled that those were the days when calls cost Rs 16 a minute, and the telephone, mobile or otherwise, was still a luxury.[26]

The second messy aspect of telecommunications policy in the 1990s lay in litigation and corruption. The murky processes by which the earliest licences were issued to provide services like email and paging were challenged in the courts. The auction of 1995 produced 'excessive and unrealistic bids' by 'small and inconsequential companies'.[27] One such company, Himachal Futuristic Communications Ltd (HFCL), based in the state of Himachal Pradesh, submitted the highest bid in nine of the 21 circles. The Minister of Communications, Sukh Ram, happened too to be from the state of Himachal Pradesh. The company had an annual turnover of Rs 2 billion, but its winning bids entailed promises of more than Rs 8.5 billion. In effect, it had more slices of birthday cake than it was able to eat. When it was clear that its bids were beyond its capacity, instead of throwing them out as the mischievous distractions of a frivolous bidder, the communications ministry, acting in its role as referee, declared that no bidder would be allowed to hold more than three licences. It gave HFCL the privilege of choosing the three licences it most fancied—in effect, getting to keep the slices of cake with the thickest icing.[28] HFCL soon sold 30 per cent of its holding to one of India's big business groups.[29]

Such deals exemplified the charges levelled by liberal economists at the controlled economy of post-1947 India: ministers defended their ministries because they and their supporters profited from them. The Communications Minister, Sukh Ram remained a force in the politics of his home state of Himachal Pradesh for fifteen years after he and the Congress government lost power in 1996. He danced nimbly in and out of the Congress Party, flirted with Bharatiya Janata Party (BJP) governments and ran his own influential state party for a time. He long avoided going to prison, though the Central Bureau of Investigation (CBI) raided his houses, seized 'disproportionate assets' and charged him with corruption in 1996, and he was convicted and sentenced to three years imprisonment in 2002 and again in 2009.[30] In January 2012, approaching 88, he was at last jailed.[31] As Minister he had, it was alleged, 'a modus vivendi with DoT [Department of Telecommunications], which earned him considerable money'.[32] In 2007, he cheerily took credit for the benefits that cheap cell phones had brought the country. 'When I took over 40 lakh [4 million] people were on the ... waiting list [for a phone] with the waiting stretching from one to 10 years'. Under his guidance the wait ended and multi-national companies were invited 'to introduce the latest technology'.[33] Analysts not connected to the ministry reached different conclusions: Sukh Ram had resisted the National Telecom Policy of 1994 which was forced on him by the Prime Minister Narasimha Rao; a minister 'would not have wished to reduce the power or influence' of his department.[34] Indeed, the department fought a delaying action that staved off the creation of an independent regulator until after the general elections of 1996. The Telecom Regulatory Authority of India (TRAI) came into being in 1997 under an act of parliament that took more than a year to pass.[35]

Act II: Sidelining the referee

In liberal economies, the theory is that independent regulators are created to take the field as referees ensuring that rules are obeyed and the public interest is served when rival teams of private enterprises compete. Until 1997, the Department of Telecommunications (DoT) not only owned the ball (the wires and the phones) and wrote the rule book (e.g., the 850 pages of *Swamy's Telephone Rules*), but blew the whistle as well (decided whether grievances had merit and action needed to be taken). The creation of TRAI was intended to take away the whistle; but the DoT still wrote the rules. And the DoT had recourse to the courts—a government department successfully challenging a government-created regulator—to get its interpretations enforced when they were at odds with the regulator's recommendations and directives.[36] Decisions of the courts in cases brought by the DoT 'reduced the role of the regulator to that of a tariff setter' in the words of the first TRAI chairman.[37] The referee was shunted to the sidelines.

In the three years between May 1996 and May 1999, India had three general elections, five governments and six Ministers of Communications.[38] At the same time, the defects of the first attempts to have private companies provide telecom services became apparent. Phones spread steadily but not spectacularly, and the government found it difficult to extract promised fees from the companies. The number of phones more than trebled in seven years—from about 5 million phones for all of India in 1991 to close to 18 million by 1998. Previously, it had taken twenty years to treble phone availability: from one million to three million between 1968 and 1987.[39] Most of the expansion in the 1990s was of landline telephones, rolled out with somewhat greater urgency by the Department of Telecommunication, now feeling pressure from its private, though hobbled, competitors. The revenue from the new mobile licensees disappointed governments, and by 1998 the private telecom companies were crying poverty, near-bankruptcy and inability to pay their dues.

The new regulator, TRAI, concluded that the cries were justified. The companies had bid too high in 1995 to succeed in a game in which the deck was stacked against them. The regulator attempted to curtail some of the advantages that the DoT enjoyed. It allowed private companies to raise their monthly rental charges from Rs 156 to Rs 600, which in turn allowed them to lower the price of calls. For the first time TRAI also gave them a share of the revenue from long-distance and overseas calls which up till then had gone entirely to the DoT, even if the call originated from the phone of another company.[40]

Such measures, however, were not enough to make the companies profitable. According to one interpretation, their collapse was precisely what the DoT wanted. It would have 'happily cashed the bank guarantees' that the private companies had deposited, seen them go out of business and 'restored its own monopolies'.[41] Politically, however, now that India was on the 'liberalization' road, no government could afford to turn back for fear of scaring off keenly wooed foreign investment.

The BJP minority government formed in March 1998 lasted thirteen months, during which time the impasse in telecom policy drove it to draft a new policy in the winter of 1998–9. The government accepted it in April 1999—and fell almost at once (though not because of telecom policy). During the period when he acted as a caretaker prior to elections in September 1999, Prime Minister Vajpayee replaced the telecommunications minister and took the ministry under his wing. Business campaigns to 'do something about the telecom mess' contributed to Vajpayee's decision. The displaced telecom minister, who advocated forcing companies to meet their dues to the government or lose their licences, complained that the private telecom businesses were being let off the hook: 'If a cricket match is going on, are the rules changed suddenly to benefit one team just because its five players get bowled out quickly?' he asked a seminar.[42] Why should they not pay the sums they had committed? Vajpayee, however, appears to have viewed the problem differently. Politically, there was pressure to accommodate businesses that were potential contributors to his party; and economically, he could see no advantage in driving the existing companies out of business—they would generate no revenue for government, nor put many phones into homes and offices, much less villages. The telecom companies thus used their 'political means', according to Desai, to 'fend off' penalisation by the government-controlled financial institutions to whom they were indebted.[43]

When the Bharatiya Janata Party formed a coalition government after the elections of September 1999, Vajpayee retained control of the telecommunications portfolio, and in October, the government began to implement the New Telecom Policy (NTP-99). The latter aimed to break deadlocks, fix defects and get loss-making private companies back on the path to profit and growth. 'The Government recognizes', NTP-99 began, 'that the result of the privatisation has so far not been entirely satisfactory'.[44]

NTP-99 forgave the old auction debts of 1995.[45] In future, operators would be required to pay government a percentage of annual revenue, not a fixed fee. At the same time, government financial institutions gave concessions to indebted telecom companies that let them escape the bankruptcy that many claimed was imminent. 'For political reasons', political parties in power could not afford to have the companies of financial contributors fail. 'Private operators learnt ... that access to political rulers was essential for survival, and could be used to neutralize' the DoT.[46]

NTP-99 aimed to make the cell phone industry profitable. It provided a twenty-year licence period; it based annual fees to government on a percentage of revenue; it reduced the charges for calls to and from the government phone provider; and it allowed private operators to offer all the services that had previously been the monopoly of the government.[47] It also called for the DoT to surrender its role as an installer of telephones to a government corporation which would do for the whole country what MTNL had been doing for New Delhi and Mumbai since 1986—install telephones, albeit in a leisurely way. This eventually happened with the creation of Bharat Sanchar Nigam Ltd (BSNL) in October 2000.[48]

The referee, TRAI, also came in for renovation. In 2000, the government gave it the unchallengeable right to make rulings about phone charges and about the terms under which one company's callers could connect to subscribers of other phone companies.[49] A new body was created to keep telecom disputes out of the courts—the Telecom Dispute Settlement and Appellate Tribunal (TDSAT). One view holds that the new telecom policy 'shifted the balance of power in favour of TRAI'.[50] In the sense that the powers of the dreaded DoT were curtailed, this is probably true. But another view holds that TRAI then fell victim to 'systematic stacking ... with retired civil servants' and became a creature of the minister of the day.[51]

A number of developments flowed from NTP-99. However irregular some of them seemed in terms of principled public policy, they produced a revolutionary expansion of mobile telephony. In 1999, India had 23 million phones; by 2005, it had close to 100 million, and more than half the subscribers now were customers, not of the government telephone provider, but of private companies. By 2012, India had 900 million mobile-phone subscribers.[52] In the three years after the endorsement of the NTP-99, telecommunications in India were transformed:

- The government changed the licence fee from a fixed sum (which companies had bid in the auction of 1995 and then proved unable to pay) to a percentage of a company's annual revenue. If you made money, you paid more; if you didn't make money, you paid less.

- Licence holders were allowed to provide every kind of telephone service without having to obtain a new licence for each service.
- Licences were granted for twenty years and were to be extendable.
- The DoT's function of installing telephones was split off into a separate government-owned corporation, Bharat Sanchar Nigam Ltd (BSNL)

In 2000, the same year that BSNL was created, the government did a deal that combined technology, politics and profit.[53] It permitted companies that had acquired the right to provide landline or 'basic' services to become—in practice, though not precisely in law—providers of *mobile* phones. The two most important beneficiaries were the Tata organisation and the Reliance group of the Ambani family which had acquired the right to provide 'basic services' in a number of cities in the auction of 1995. Only a few companies had bid to provide 'basic services' because this involved laying (expensive) cable and competing with the DoT's existing landline operation which had hitherto been the only service available. The companies that opted to provide such services were allowed to complete the 'last mile', or the connection from base stations to individual houses and businesses, using a wireless technology favoured in the US to operate mobile-phone networks.[54]

The Tata and Reliance companies laid cables that carried calls to base stations in the cities in which they operated. From there, calls were carried to homes and businesses—the 'last mile' in the jargon of telecom companies—by wireless, using CDMA technology, to handsets that were cordless. In theory, such a setup was meant to behave like a phone call to a household cordless phone: you took the call in your house or office. In practice, however, the service provider could—and did!—give you the capacity of a mobile phone: your calls could be made and connected anywhere in the urban area that the phone company had its base stations. It was the mobile phone you had when you did not, strictly speaking, have a mobile phone. Users got a cell phone much more cheaply than if they had subscribed to a properly licensed provider because their supplier had paid much less for a 'basic service' licence than companies that were providing fully licensed *mobile-phone* services.[55] Stories abound of doctors, dentists and ordinary citizens carrying cordless phones the size of loaves of bread on their daily rounds in Kolkata, Mumbai and other cities.

Companies that had paid for the privilege of providing 'legitimate' mobile phone connections shrieked of foul play and went to the newly created Telecom Dispute Settlement and Appellate Tribunal (TDSAT). Two years of legal pass-the-parcel followed: decisions of TDSAT were appealed to the Supreme Court, which sent cases back to TDSAT which returned them to TRAI. The outcome in November 2003 was that that every private company with any sort of telecom licence was given the chance to provide all telecom services. Those that had bought only 'basic service' licences had to pay a premium to put them on a more level footing with those that had paid heavily for 'genuine' mobile licences. The Ambani family's Reliance group was fined for having offered mobile-phone services without having a mobile-phone licence.[56]

The scandals and mismanagement that characterise the expansion of telecommunications were not unique to India, though they often had a distinctively Indian flavour. In the US, the auctions and lotteries of the 1980s meant that 'any chancer [speculative investor] could land a ten-year licence, then hawk it around [to] more competent operators'. The result was an 'extremely disjointed' system in which 'roaming'—using a phone in different regions—was 'extremely difficult', and the US by the mid-1990s had 'a crazy-paving of licences covering the country'.[57]

Europe did better, and by 1995, the year mobile phones were first sold in India, 'European coverage was nearly complete'.[58] Later, some European countries made policy blunders similar to India's. When they moved to a higher level of mobile communications—so-called 3G or third generation technology—they auctioned small chunks of spectrum for huge sums. But the amounts 'placed the winning bidders in financial difficulty so that investments ...have been delayed' and the licensed frequencies were used 'very inefficiently',[59] a result similar to the outcome of Indian auctions in the 1990s.

Management of radio frequency was intensely political. It involved governments and private enterprise, huge investments and technical judgements—some would say, gambles. Private, public and national interests confronted each other. India was not alone in tying itself in knots over telecom policy. But India brought unique features to complex problems of policy-making, as the telecom scandals of the twenty-first century made clear.

Act III: Bread, clothing, shelter—and a mobile

Act III of this story began in 2003 as the telecom industry settled down to the business of selling phones and making money. In one version, the Wicked Witch of government bureaucracy lost most of her powers, her subjects were able to speak and a few handsome princes, otherwise known as private telecom companies, gave the people miraculous ways of communicating. In another version, an impotent old King (read: a decaying pretence of socialism) lost his ability to care for his people, and self-interested young princes plotted to carve up the kingdom for private gain. More mundanely, telecommunications came to be dominated by some of the great names of Indian capitalism and influenced by some of India's most notable figures of politics and government.

There were no rags-to-riches stories at this top level of telecommunications. In 2010, as a landmark scandal matured, eight major entities, two of them government-owned, had emerged as the controllers of the cell-phone industry. It needed too much investment and influence with governments for any but India's biggest players and their foreign partners to start up and survive as providers of telecom services.

Bharti Airtel held the largest share of the business with about 22 per cent of all telephones in the country using its services.[60] One source noted that Sunil Mittal (b. 1957), the founder of the company, started it with US $900 in 1995. But Mittal came from a Marwari commercial tradition that had exercised immense influence on India's economy. From his family's base in Punjab, his father had been well enough connected to be installed in the upper house of the Indian parliament by the Congress Party. Mittal was in business from the age of eighteen, first overseeing the manufacture of bicycle parts in his hometown of Ludhiana, whose commercial middle class has been celebrated and sent up in Pankaj Misra's *Butter Chicken in Ludhiana* and Anand Giridharadas' *India Calling*.[61] Mittal moved into the import of electric generators in the early 1980s, and when such imports got stifled by government bans in India's bad old, licence-permit days, he switched to making telephones, once the government allowed private companies into that business in 1984.[62] 'The romance of Bharti [the company] with telecom started during [Rajiv Gandhi's] leadership', Sunil Mittal told a US audience twenty years later.[63] The company worked with foreign firms, attracted foreign investment, including from Singapore's Singtel, and floated on the Mumbai Stock Exchange in 2002 to raise funds for a capital-hungry industry, particularly in its starting stages.[64] In 2010, Airtel went into Africa when it took over a Kuwaiti company, Zain, and expanded its investments in Bangladesh and Sri Lanka.[65]

The families behind the second and third largest telecom providers in 2011 had remarkable similarities with Sunil Mittal's family. Reliance Telecom, one of the companies of Anil Ambani, was the biggest of the cell-phone providers that used CDMA technology and fifth among those who used GSM. The Ambani brothers, Mukesh and Anil, were heirs of a buccaneering father, Dhirubhai Ambani. He came out of a small town called Chorwad ('thieves-den', according to some translators' version) on the Gujarat coast and built the biggest industrial house in India by the time he died in 2002. Reliance's success in the mobile-phone business was, its critics would say, completely in character: the circumstances were murky. Reliance acquired 'basic service' licences in the auctions of 1995. Basic licences were much cheaper than licences to provide mobile phones. Reliance adopted CDMA, an efficient but less widespread US-based technology, to provide its subcribers with the 'last mile' connection between Reliance exchanges and individual households and businesses, thus eliminating the expensive need to run a wire to every telephone. Reliance then began, illegally, to enable its subscribers to use their cordless household and office phones as mobile phones: they could send and receive calls anywhere within reach of Reliance base stations. Reliance developed a further imaginative technique: it routed incoming international calls through domestic phone numbers to eliminate the need to pay fees to India's international call company.[66]

The all-out entry of Reliance, perhaps the best known company name in India, accelerated the rapidly growing awareness and popularity of the cell phone. Reliance used a sentiment attributed to Dhirubhai Ambani: 'Make a phone call cheaper than a postcard and you will usher in a revolutionary transformation in the lives of millions of Indians'.[67] The Reliance marketing slogans presented the mobile phone as something everyone could afford and needed to own, particularly the not-so-literate: *'Chitti likhane ka zamana gaya'*—gone are the days of writing letters. Another slogan took the old political standby of bread, clothing and shelter and added 'mobile'—*'Roti, kapda, makaan aur mobile'*.[68] By 2010, Reliance was the undisputed leader in providing mobile telephones through CDMA with 53 per cent of CDMA subscribers, though with fewer than 10 per cent of GSM subscribers.[69]

The third of the biggest mobile-service providers was Vodafone Essar, usually referred to as Vodafone. The ambiguity over the name highlighted the essence of the company: this was the new India, and foreign investors were welcome. The UK-based Vodafone became a two-thirds shareholder in 2007, but Essar had been in the mobile-phone business since 1994 when it got a licence for New Delhi in the allocation of licences for the four great cities (Mumbai, Chennai, Kolkata and New Delhi). Essar was 'old India' in that it was a family firm originating from Marwar and controlled by two brothers, Shashi Ruia and Ravi Ruia, rated No. 245 in the world on the *Forbes* magazine 'rich list'.[70] Formed in 1969, Essar had grown in the days of India's controlled economy, but quickly adapted to the reduction of economic controls after 1991. It went from construction, shipping, oil-drilling and steel-making into a partnership in telecommunications with Hutchison Whampoa, the Hong Kong-based group, which sold to Vodafone in 2007. By 2010, Vodafone had about 17 per cent of the total mobile phone business in India.[71] (See Illus. 5).

India's two most famous business families of the twentieth century, the Tatas and the Birlas, could not ignore telecommunications and the mobile phone. Tata adopted CDMA technology, and like Reliance, benefited from the ability of being able to offer limited mobile service for cheap landline prices in the anarchic 1999–2002 period.[72] It emerged as the second-ranked CDMA provider after Reliance. Overall, including GSM and CDMA technologies, Tata Telecommunications ranked with the government provider, BSNL, and the Aditya Birla group's Idea, each commanding between 10 and 12 per cent of the mobile-phone market in 2010.[73] Birla's Idea had acquired licences in 1995, and through mergers (with Spice, a venture of another Marwari family, the Modis, in 2008) and partnerships (with Telekom International Malaysia), Idea could claim to be the third largest revenue earner in the mobile phone business.[74]

The list thus far looked like a portrait gallery of established Indian capitalists: Tata, Birla, Ambani, Ruia, Mittal, all from western India and from the classic Marwari-Bania-Parsi backgrounds that typified Indian capitalism as it developed in the twentieth century. The odd company in a list of dominant players was Aircel, market leader in Tamil Nadu, in which the Reddy family, owners of Apollo Hospitals, were minority partners. Dr Pratap Reddy returned from the US and established what became India's premier chain of private hospitals in 1983. In 2006, in partnership with Maxis Communications of Malaysia, they took over Aircel. One of Pratap Reddy's four daughters explained the reasons:

> We believe that telecom is one of the fastest-growing service sectors in India. Like the healthcare sector, telecom also helps us to reach out to the masses in a big way... We are testing the waters.[75]

Suneeta Reddy identified a key fact: this was technology for the masses. Potentially, the mobile phone could be as essential to an individual's sense of well-being as health itself.

India's great capitalists understood that telecommunications and the mobile phone were something big, even if no one knew where the technology was going or from where the profits would necessarily come. In retrospect, initial estimates of the potential of the mobile phone in India had been laughably modest. The companies that began to set up mobile networks in the four big metros in 1995 aimed to serve about 20,000 subscribers in Mumbai, 15,000 in New Delhi, 5,000 in Chennai and 3,000 in Kolkata. When Essar, which had won a New Delhi licence, rolled out a network to accommodate 100,000 subscribers, rivals laughed.[76]

The telephone needs of New Delhi and Mumbai had been provided since 1986 by MTNL, the government corporation created in the enthusiastic days of Rajiv Gandhi's new government and his commitment to improve Indian technology. MTNL was intended to be the forerunner of telecoms cut free from monopoly control by a weary government. It was only in 2000, however, that the remainder of the government's telephone investment that covered the rest of India was turned into the government-owned Bharat Sanchar Nigam Ltd (BSNL). Its 350,000 employees, according to one critic, were 'not formally trained to be customer friendly, and the tag of an uncaring company stuck to BSNL ...'[77] The company boasted that it had the most extensive coverage in

India: 'Whether it is inaccessible areas of Siachen glacier and North-eastern region of the country, BSNL serves its customers ...'[78] Customers, however, did not share the sentiment. BSNL came to be known as *Bhaai Saahib, Nahin Lagega*— 'Brother, it won't be connected'.[79] BSNL's share of mobile-phone subscribers fell from almost 20 per cent in 2006, when its widespread infrastructure gave it advantage, to just over 12 per cent by 2011, by which time private competitors had built towers, laid cable and advertised relentlessly.[80]

Act IV: Schools for Scandal

'The present seems so nineteenth century', wrote the historian Richard White as he compared the frenzied enthusiasm for the Internet in 2011 with railway expansion 150 years ago. Both involved 'public' goods—land grants for railways, Radio Frequency allocation for telecommunications—vast sums of money and breath-taking corruption and political intrigue. At one point, an American railway president could marvel at the fact that he had 'two Senators of the United States—one of them the acting Vice President' ready to sell him their votes on a piece of railway legislation for $25,000.[81] The twists and turns in India's cell-phone story were not unique.

The scandals that originated in 2007 illustrated the way in which officials, corporations and politicians could benefit each other through collusion in the distribution of public property—in this case Radio Frequency.[82] Like the American railroads of the nineteenth century, India's telecom giants 'thoroughly insinuated themselves into the modern state. Through their lobbies and friendships, they could be found in [the United States] Congress, the legislatures, the bureaucracies and the courts'.[83] In similar ways, key players in the allocation of spectrum for second-generation telecommunications (2G) engineered improprieties in 2007 and 2008 that dwarfed the escapades of the 1994–5 auctions. The results would percolate through Indian politics and judicial systems for years. But by February 2011, the former Minister of Communications, A. Raja, had been arrested for corruption and held in jail, though not tried or convicted. Allegations about who knew what and when touched even the upright Prime Minister Manmohan Singh and tainted various politicians and television journalists. So blatant were the breaches of process and administrative rectitude that the Central Bureau of Intelligence (CBI), India's national crime investigator, did something never done before: it raided a government ministry—the Ministry of Communications—to seize documents thought to provide evidence of corruption.

In 2007, with sales of mobile phones skyrocketing and the industry flourishing, the government announced it would licence the use of additional spectrum, previously reserved for national interests. More spectrum meant wider Radio Frequency roadways which allowed more elaborate services—video, music and so on—and enabled more voice calls to be made simultaneously. In a booming mobile-phone market, additional spectrum was immensely valuable. However, it was neither auctioned to the highest bidders, nor allotted on the basis of an applicant's proven competence. Rather, it was doled out on a spurious first-come-first-served basis. The price was unrealistically low: applicants paid the price that spectrum had fetched in 2001. This took no account of inflation between 2001 and 2007 or that in 2001 mobile telephony was in its infancy in India. But even more remarkable—comical had it not been so cynical and serious—was the deadline for payment and receipt of Letters of Intent (LIs) from the Department of Telecommunications. 'After keeping ... applicants on tenterhooks for almost three months', on 10 January 2008, the Department 'announced at 2.45 pm through a press release that LIs would be issued between 3.30 and 4.30 p.m. giving a window of only one hour'. Thomas K. Thomas, an experienced telecom journalist, described 'total pandemonium' with one company employing 'a hired goon to ensure that it was ... first to enter the gates'. One of the biggest winners was a real estate company called Unitech with little telecom experience; it gained licences for all twenty-two telecom circles or jurisdictions and soon sold more than 60 per cent to Telenor of Norway for four times what it had paid.[84]

By this process, spectrum was sold for bargain prices: a licence for the right to Radio Frequency covering the whole of India went for Rs. 1,651 crores (about US $300 million). The later inquiry by the Comptroller and Auditor General (CAG) estimated that an open auction could have raised a total sum of up to Rs 176,000 crores (i.e., 176,000 multiplied by 10 million) or close to US $40 billion.[85] The CAG report emphasised the spectacular disregard for rules, procedure and fairness. At short notice, with insiders apparently informed in advance, 122 licences were issued on a single day.[86] 'The entire process ... lacked transparency and objectivity'. The CAG report continued: 'Eighty five out of the 122 licenses issued in 2008 were found to be issued to Companies which did not satisfy the basic eligibility conditions set by the DoT'.[87] None of the thirteen companies that had been awarded the eighty-five licences had the paid-up capital required to qualify for a licence—in short, they were not large enough to finance the enterprise they proposed to create, and their claim to have such capital was 'false and fictitious'.[88] Although the timing of receiving Letters of Intent was made known only hours in advance, '13 applications were ... ready with Demand Drafts' drawn earlier—evidence, in the view of the Auditor-General, that someone had been 'selectively leaking' the crucial date and time.[89] A number of the lucky licensees, who had little or no experience in telecommunications, quickly sold chunks of their holdings and made substantial gains. The Auditor-General's report estimated that if the spectrum had been auctioned, it would have brought the government about five times the revenue that in fact received.[90]

More remarkable, the impugned Minister for Communications, A. Raja, and his department had disregarded written advice not only from the TRAI but from the Prime Minister himself. In January 2011, Kapil Sibal, the minister who replaced Raja, tried unsuccessfully to fend off the creation of a Joint Parliamentary Committee to probe the spectrum scandal. Sibal argued that the Prime Minister never actually differed with Raja over the policy for allocating spectrum. When Manmohan Singh wrote to the Communications Ministry about spectrum allocation in November 2007, the Prime Minister was, according to Sibal, only

> forwarding a summary of the suggestions received and requested the Minister to consider all these aspects carefully. This has been wrongly interpreted in the CAG report as a direction to act in a particular manner on the issue of pricing. The only direction given by the Prime Minister was that the Ministry should act in a fair and transparent manner and keep the PM informed.

The Ministry replied with justifications for its actions, which, according to Sibal, the 'Prime Minister accepted'.[91] A less sympathetic interpretation was that Raja and his ministry had brushed off a prime ministerial letter, confident that they could get away with it. For a time they did, though Raja eventually found himself charged and in jail for more than a year after his arrest in February 2011.

In the long term, the scandal and its consequences may have forced the consolidation of the telecommunications industry. In February 2012, the Supreme Court revoked all 122 licences issues under the flawed process of 2008. It said, in effect, 'Start again. Re-auction the spectrum. Companies will surrender spectrum that was illegally acquired'.[92] The decision threw such companies, particularly those that had built networks and enrolled customers, into grim uncertainty. The circumstances presented various choices. Foreign companies could quit India and advise others to avoid India as a place for investment; Indian companies, and foreign companies untouched by the Supreme Court judgement, could welcome a reduction in the number of competitors and a chance to bid anew in whatever auction process governments might devise. By world standards, the fifteen telecom companies India supported in 2011 were more than most countries could bear.

The '2G scandal' had a further dimension. It brought together two opposite aspects of the cell phone: the immense value of Radio Frequency to great corporations and the cheap everydayness of the mobile. Its taken-for-granted quality led people to use their mobiles as if they were talking privately to someone face-to-face. A spectacular fallout of the '2G Scandal' was that conversations of a highly sophisticated but technologically naïve lobbyist, which had been recorded by the income tax department, were leaked to the media. The lobbyist had worked hard to see that Raja got the telecom ministry in the coalition government formed after the Congress Party victory in May 2009. She had talked regularly and all too candidly to television journalists, politicians and business people—on her mobile phone. Everyone involved seemed to think such conversations were private and secure.[93] In 2010, they became national public property (Chapter 8). The vast sums involved, and the blatant bartering of influence, threatened the coalition government, since the disgraced Minister of Communications was a member of the DMK, the second-largest party in the national coalition government after the Congress.

The scandals publicised what India's great business conglomerates had known for twenty years: vast sums were to be made from this invisible, intangible thing called Radio Frequency and because its allocation was in the hands of governments, it was crucial to have sympathetic politicians in charge of key departments. The scandals also underlined the way in which the mobile phone in less than ten years had become an object as familiar to hundreds of millions of people as a pair of sandals. However, as the protagonists whose phones were tapped discovered, it was easier to track mobile phone calls than footprints; records of phone calls did not get washed away in the first shower.

When Raja was being held in New Delhi's legendary Tihar Jail in 2011, his patron and political leader, K. Karunanidhi, chief minister of the state of Tamil Nadu, praised him as a man 'languishing in jail and paying the price for making mobile phones affordable to the people'.[94] It was the same attempt to win sympathy that Sukh Ram, the Communications Minister during the 1994–5 auction fiascos, had revelled in. Given the affection that voters in the switched-on state of Tamil Nadu had for their mobile phones, it was an ingenious ploy.

Contests over allocation of Radio Frequency spectrum, and graft arising from such contests, will continue as long as consumers show a passion to communicate in ever more elaborate and spectrum-hungry ways. Technicians will search for ways of packing bigger digital-signal loads onto radio-frequency roadways. And the people who run corporations and governments will struggle to broaden and dominate these roadways for the wealth that such control can generate. These contests went on among relatively few entrepreneurs, bureaucrats, politicians and technicians. Closer to the base of the Indian pyramid, what people took for granted like the blue sky behind a summer haze was that cheap cell phones were part of life in the first decade of the twenty-first century. To hundreds of millions of users, it did not matter which corporation provided the service or which politicians carved themselves a corrupt slice of the action. What was important was that there was a cell-phone service and that it was cheap and reliable.

'The telecom industry', Sunil Mittal asserted in 2005, 'is now employing 3.7 million people … The media industry, the advertising industry, the content industry, all are riding on the back of telecom …'[95] It was true that a cascade of occupations grew up to serve a new industry. Women and men who diced with billions and lobbied politicians; Mumbai advertising agents and marketers who devised campaigns to sell phones; technicians who wrote software and positioned cell-phone towers; mechanics who built towers and property owners on whose land towers were built; distributors and shopkeepers who put phones in front of customers. And customers themselves—who found a myriad of uses for what once had been a squat, black, wire-connected, bakelite block locked in a box in the room of a social superior and good only for talking into. How these various interests, skills and initiatives connected with one another, and began to change practices and expectations, are the subject of the next chapters.

PART TWO

CONNECTING

3

MISSIONARIES OF THE MOBILE

We are the most important in the chain!
Sumit Churasia, Nokia sales promoter, Banaras, 2009

'Speak directly into the mouthpiece, keeping moustache out of opening', advised an advertisement for the telephone in California in 1884.[1] People had to be educated about how to use a phone and persuaded that they might want to use one. Similar education was necessary for the mobile phone in India. In old societies, as we saw in Chapter 1, information reinforced power, and such relationships carried over into independent India when phones were rare and locked away. Under the government monopoly, a telephone was a privilege that a citizen had to demonstrate a right to own. In the monumental *Swamy's Treatise on Telephone Rules*, the section relating to 'New Telephone Connections' ran to twenty-two pages. 'Every aspirant for a permanent telephone connection' was required to 'apply in the prescribed form', accompany it with Rs 10 non-refundable fee and be reassured that their application was 'valid for one year from the date of issue'.

Once chunks of radio frequency spectrum were sold to private operators in 1994, the operators had to find ways to get their money back, either by selling expensive services to thousands of prosperous customers or inexpensive services to millions. As we saw in Chapter 2, it took ten years for the entrepreneurs and the state to come up with a formula that allowed money to be made and mobile phone usage to mushroom. During that time, people learned. They learned through advertising campaigns; they learned through sales techniques that responded to their needs and price preferences; and they learned through the creation of vast networks of vendors who were trained well enough in the mechanics of mobile phones to demonstrate them and set them up for others. This chapter outlines how the missionaries of the mobile took the phone to the people.

Man's best friend

Seminal advertising campaigns, financed by great corporations, spread the word about the wonders of the cell phone across India. Marketing departments trained thousands of sellers and sales promoters who then explained to ten of millions that the device was simple, useful and economical. In time, these sellers and agents communicated the key requirements of customers back to manufacturers and service providers, and this process shaped the way mobile phones were designed and their uses proclaimed to potential buyers.

When Airtel, India's largest provider of mobile phone services, changed its advertising agency in August 2011, it was news only briefly on the business pages of Indian newspapers. But the funds involved in Airtel's advertising account were huge: Rs 400 crores (about US $80 million) annually by some estimates, and in the advertising world, the switch was depicted as a dire day for Rediffusion, the agency that lost the Airtel account.[2] The switch from Rediffusion to JWT (J. Walter Thompson) ended a relationship that began when Airtel started in the mobile-phone business in 1995. For fifteen years, the two companies had worked together to create strategies and carry out advertising campaigns costing hundreds of millions of dollars (hundreds of crores of rupees) and involving some of India's greatest celebrities including the cricketer Sachin Tendulkar and the multi award-winning musician A. R. Rahman.

The campaigns illustrated both the difficulties of trying to 'know India' and the shifting circumstances of the telecommunications industry. As we have seen, in the ten years after 1991, as mobile telephony was opened to private enterprise, scandals and obstacles were spectacular but the growth in paying subscribers was not. In 1991, there were five million phones in India, none of them mobiles; by 2001, there were 36 million phones, but fewer than 4 million were mobiles.[3] It was not a good decade to have been selling mobile phone connections.

By launching its mobile service in Delhi in 1995, Airtel acknowledged the widely held preconception that the new device was geared for the educated and wealthy, present in larger proportions in the national capital than any other city. Early advertising campaigns reflected this expectation. Skewed towards television, they emphasised the business advantages to be gained from having a mobile phone. Cost was prohibitive for anyone except the well-to-do: calls in 1999 were charged at Rs 16 a minute (about US 30 cents) when Rs 16 was half a day's wage for a labourer.[4]

In the early decades of the telephone in the US, the dominant Bell company had invested heavily in trying to expand its market beyond businesses and to educate people about the different ways they might use the phone. Yet even Bell was 'slow to employ as a sales tool' the use that later came to dominate: 'sociable conversation'.[5] There was even less temptation in India to try to market idle chatter as a reason for wanting a mobile phone: the costs of the phone and its calls were too great. In any case, the phone still suffered from the elitist hangover of the colonial and Nehru eras: phones were meant for serious people and important matters.

The breaking of the cost barrier happened in 2003 when a minute of talk time on a mobile phone fell below two rupees (one US cent), and advertising men and women punctured much of the telephone's pompous solemnity. When the industry began to expand after the amnesties and concessions by government in 2000–02, Hutchison Essar, a joint-venture between the Indian Ruia group and Hutchison-Whampoa of Hong Kong, developed an advertising campaign designed to domesticate the mobile phone. For this purpose they used Cheeka, a small dog, who soon became the celebrated figure of one of the most popular campaigns in Indian advertising history. The ability to measure the effect of advertising on sales has eluded marketers for a hundred years, but the Hutch dog ads might claim some credit for the growth of the company's subscribers from 1.8 million in 2002 to 15.1 million by 2006 and a 17 per cent share of the mobile phone market.[6]

The dog-and-phone advertisement, created for television and adapted for print, reflected the way advertisers envisaged the market for cell phones in the mushroom years after 2002. The targets were urban people who did not need instruction about how to operate a phone or why it might be useful. The visuals involved small children doing endearing things, always followed by an even smaller dog—a pug—and the message was that the coverage, like the dog, followed you everywhere: 'Wherever you go, our network follows'.

The Cheeka campaign illustrated the cascade of occupations, extending from Mumbai-based advertising professionals to small-shop entrepreneurs that grew up around the mobile phone. Once the telecom industry was unleashed with the bargain-basement Universal Access Service Licences (UASL) of 2003, half a dozen private companies competed fiercely and dragged the two government-owned companies, MTNL and BSNL, into a more competitive environment as well. The Hutchison-Essar group gave its advertising account to the international firm Ogilvy and Mather in Mumbai. The first Hutch advertisements featuring Cheeka the dog were broadcast in September 2003.

During an all-night brainstorm came the idea about how to tell a story of reliability, simplicity and mobility 'without the tech, using emotions'.[7] The creators of the advertisement initially toyed with the idea of having a small girl following her brother, but decided 'a dog would be less mushy'.[8] The television campaign ran to more than a dozen ads in which the dog followed a small boy everywhere: bathing, getting a haircut, going to school. 'Man's best friend was the simplest analogy that could represent unconditional support', an Ogilvy and Mather executive explained.[9] Kapil Arora, a senior vice-president, recounted how a pug dog became a brand representative for mobile telephony:

> Cheeka herself was a happy accident. We were looking at using another dog. But at the [television] shoot in Goa the dog refused to perform. ... It was raining and the team was completely desperate because it had to finish something, so the production crew ran around looking for dogs and they found this sweet dog there who took an instant liking to the boy and started following him around. It was a happy coincidence that we happened to use the pug.[10]

The campaign went from television into print and outdoor advertising (including stick-on paw marks put on floors leading to elevators in large office buildings) and won an award as the top print advertisement of 2003.[11] Eight years later Cheeka, had a

Wikipedia entry, was the subject of a chapter in a book and was judged to have 'made Indian advertising history'.[12] (See Illus. 6). Much of the expenditure in the Hutchison advertising campaign—an estimated US $40 million—went into the purchase of television time, newspaper space and the printing and dissemination of billboards and displays for stores and shops. When Hutchison sold its stake in the company to Vodafone in 2007, Cheeka was part of the campaign to educate consumers into a new corporate name and colour: she left a pink (Hutchison colour) doghouse and returned to an improved red (Vodafone) one. In the print advertisement and outdoor hoardings and billboards, Cheeka appeared in a red kennel.[13] The transition to the new brand name was touted as 'the largest brand change ever undertaken' in India and 'as big as any in the world'. The extent underlined the range of people becoming enmeshed in the mobile phone industry: 400,000 retail outlets, 350 Vodafone-only stores, 1,000 smaller stores and an advertising cost estimated at US $40 million. To advertise the change, Vodafone bought all the advertising time on fourteen channels of the Star Network for a full day. Only the 'Hutch turns to Vodafone' advertisements ran. Later research was said to have shown that 80 per cent of the viewing audience had got the message. By 2010, Vodafone had more than 100 million subscribers and was one of the three largest mobile service providers in India.[14]

Why should a pug dog born in Britain have been seen as an effective sales symbol for mobile phones in India?[15] As Rahul Singh points out, dogs 'terrify many Indians', are regarded as dirty by others and help to give India the highest number of rabies deaths in the world.[16] 'In India, man's best friend often is not', a front-page headline in the *International Herald Tribune* advised readers, and reported on dangerous 'packs of strays' roaming towns and cities and causing an estimated 20,000 rabies deaths a year.[17] Why a dog? Social class, and changes in class composition, provide part of an answer. The initial Cheeka campaign was aimed at television viewers, and in 2003, India had about 95 million television households in a population of 1.03 billion people. If those sets had been distributed evenly, about 45 per cent of the population would have slept each night in a place with a TV set. These were the targets of mobile-phone advertising.[18] The first barrier lay in capturing their attention, the second, in convincing them of the simplicity and steadfastness of the device.

Marketing of goods in India has always presented intriguing problems and solutions.[19] William Mazzarella devoted two chapters of a book about Indian advertising to trying to understand the development of a single campaign to market mobile phones in Mumbai in the mid-1990s. The essence of success, it appeared, was to create messages in which 'Indianness' looked like 'an avatara [incarnation] of the global'.[20] When the Reliance group of the Ambani family moved into mobile phones in 1998 through Reliance Telecom Limited (RTL), it faced all the old problems of marketers—distribution, social conservatism, customer frugality—and, as well, the need to educate purchasers in a technology that alarmed many consumers even in urban and industrialised countries. Reliance deployed 'massive aggressive promotion campaigns, using every possible medium of communication',[21] which since the 1980s had proved effective in launching new products, including newspapers.[22] As well as the publicity push and enticingly cheap plans to attract subscribers, Reliance attempted to set up a missionary society of sales representatives—50,000 Dhirubhai Ambani Entrepreneurs (DAEs). Intended to capture people in the streets and sell them mobile phones, the 'entrepreneurs' had to pay Rs 10,000 as a deposit but were promised Rs 100 for every subscriber they enrolled. The scheme flopped, partly because the phone services did not initially fulfil the promises, but also because the 'entrepreneurs' had not been sufficiently trained and confident in what they were selling.

Talk time—small, medium, large

The Ambani name and Reliance brand attracted a host of small commercial people who would have happily tried selling woolly mittens in Mumbai if the name 'Reliance' had been emblazoned on them. What was learned, however, was the need for training. Enthusiasm and commercial experience were not enough. There had to be a basic understanding of the technology, an ability to interpret it to customers and the appearance that agents knew what they were talking about. 'Consumers found it difficult', an *Outlook* magazine team concluded, 'to trust paanwallahs and neighbourhood grocers with their cash and post-dated cheques'. A prospective purchaser identified what was missing: 'All [the agent] could say was: *"Yeh bahut cheap hai"*(This is very cheap)'.[23]

Agents did not understand how the phone worked or how payment was to be made. (See Illus. 7).

Reliance's campaign to create 'barefoot mobile sales reps' did not work at first, but it found that existing shops could be made into effective sales outlets for many of its products. Its aggressive rival, Vodafone, had 600,000 'points of presence' when it took over Hutch in 2007—and 1.2 million by 2010.[24] The crucial ingredient in these transactions was simplicity. People wanted something that they could both afford and operate. To be effective, sales agents needed something that was easily understood and explained, and they needed to see profit in handling a new and initially mysterious item. Established shopkeepers and merchants, especially if they had offspring unafraid of the technology, took on dealerships as part of existing businesses. Anuradha Aggarwal, a senior vice-president of Vodafone, spent thirteen years as an executive in FMCG industries (Fast Moving Consumer Goods) in Uttar Pradesh and Maharashtra. In 2001, Saharanpur, a town in western Uttar Pradesh, served a district of about 3 million people; it had no stores selling telecommunications products. In 2011, the district population had grown to 3.5 million, and 40 per cent of the shops in Saharanpur's main market did some sort of commerce related to mobile phones. 'I keep telling these guys', Aggarwal said of conversations with marketers of mobile phones, '"You have not seen what a tough patch is, because the consumer wants your product"'.[25] Commercial sense and listening to customers told shopkeepers that mobile phones and mobile-phone services excited curiosity and tempted people to spend money. But ease with the technology was essential to satisfy customers and turn them into enthusiasts telling their friends and relations about the benefits of mobile phones and returning regularly to replenish talk-time and sample other products.

Probing for ways to turn a profit, the providers of mobile phone services found that the snail-paced landline model was dead. Just as cash-strapped customers needed to buy matches, soap and shampoo quickly and in small quantities, so they wished to purchase their mobile-phone time. To keep within their budgets, customers wanted to pay for their calls in small amounts in advance. In addition, the pre-paid mechanism made service more sociable. Hesitant people when buying a pre-paid connection were assured that a shopkeeper they knew would carry out the recharge process for them. Consumers left the shop with their mobiles ready for use. And if anything went wrong, there was a familiar shopkeeper to go back to. Combining the economic and the social gave the pre-paid card a powerful attraction. In 1999, about a quarter of mobile phone subscribers had pre-paid connections; by 2002, this had risen to two-thirds.[26] Price was crucial, as the careful budgeting of poor people made clear, and the cut-throat competition that resulted cut the cost of calls and phones. Companies struggled to reshape the industry to their own advantage as two examples illustrate.

Bharti Airtel emerged as one of the vigorous survivors of the chaotic first five years of mobile telephony. It started its pre-paid cell-phone card under the name of 'Magic' in January 1999 in Delhi, where the company had built its initial base. Airtel's marketing people aimed to provide a breath-taking contrast to earlier experiences of trying to get a phone connection. With the Magic pre-paid card, purchasers walked out of the store in half an hour with working phones. Everything was intended to be cheap, easy and fast. An agent explained to the customer how to make the phone work, recharge the card, and keep costs within the customer's budget. With a pre-paid plan, the phone stopped working when you had used up the time you had paid for.

Tens of millions of people had to acquire telephones if mobile telephony was to reward the huge investment required to build and maintain vast networks of cell-phone towers. No company could afford to open thousands of sales outlets that sold only mobile phone services. Airtel and its competitors therefore began to incorporate all kinds of retail shops into their sales chain, including the tiny general stores found in every small town where cigarettes were sold individually and shampoo could be bought in one-wash packets. Smalls shops in cities, towns and villages had long been part of the networks built by Indian marketers of FMCG—Fast Moving Consumer Goods. These outlets became the targets of the great capitalists selling mobile phones and services. The aim was to make cell-phone services available even to those who could make only a small investment. The initial Airtel Magic packages could be bought for as little Rs 300 (though this represented more than a week's wages for an agricultural labourer in 2002). 'By 2002', Radhika and Mukund write, the Airtel Magic card 'became the largest selling pre-paid cellular card in India' with 6.4 million subscribers. The number had doubled within a year, and pre-paid cards were already generating more than half of the company's revenue.[27] Ten years later, when Indian cell phone subscribers exceeded 800 million, the numbers of 2002 seemed paltry. But in their time, they constituted an immense breakthrough—a stunning contrast to the old experience of applying for a landline and waiting two years for a connection.

Such a breakthrough required connecting with, and educating, the shopkeepers. Fiercely competitive service providers planned the search for customers like military operations. 'I was with Vodafone earlier in T[amil] N[adu]', Bobby Sebastian, a mobile-phone veteran, explained in 2010.

> We had mapped our distribution network ... We used to take lat-long [latitude-longitude] of each of these [shop locations] ... I have ... a distributor [who] ... has got some 200 shops with him. [We] go and take a lat-long—we will map it onto Google Earth. By mapping it to Google Earth I clearly know how my distribution is. [Our distributor] might have covered only one area. So with the help of Google Earth, we used to set up distribution in the areas where we have not got it. Suddenly I will find a village where I don't have a retail outlet. I'll ask my distributor to go and cover that area ... So the reach and the visibility ... penetrated to that extent.[28]

To sell recharge coupons required little more expertise than selling packets of shampoo; but to sell SIM (Subscriber Identity Module) cards, the thumb-nail-sized piece of circuitry that slides into a phone to store information and enable the phone to connect to the Radio Frequency of the company supplying the service, meant that the seller would often be asked to install the card. Not surprisingly, SIM card sellers were fewer; but even so, the number was remarkable. At one point in 2009, Sebastian said, Vodafone had 70,000 retailers selling recharge coupons and 40,000 activating SIM cards in his region alone in Tamil Nadu state. At an all-India level, Vodafone had 600,000 'points of presence' in 2007, and 1.2 million by 2010.[29]

For the customer, the transaction of buying a SIM card disguised the complicated procedures that lay behind simplicity. Tens of thousands of dealers had to be trained to sell and install SIM cards. A customer who already owned a phone and wanted to sign up for Airtel service went to a dealer displaying the Magic logo and bought a Magic SIM card which the dealer inserted into the phone. To activate the phone, the dealer sent the unique numerical details of the SIM card to the Magic network by SMS, and the customer had a connection. To the uninitiated, the procedure seemed difficult, though it took only a few experiments with phones to learn how to change SIM cards confidently. Nevertheless, it took thousands of hours of training to bring tens of thousands of distributors, travelling salespersons and small shopkeepers to a basic level of confidence where they were able to do more for customers than say, *'Yeh bahut cheap hai'*.[30]

While Bharti Airtel was focussing its efforts on promoting the pre-paid SIM revolution, the Ambani family's Reliance and Tata Teleservices, both of which had acquired landline licences—'basic service' licences, as they were called—at low rates, sent the licensees of mobile phones into a frenzy by offering clients a mobile service at a landline rate in 2001. This was the 'loaf of bread' cordless phone referred to in Chapter 2. The companies supplied what was intended to be an old-fashioned, wire-connected telephone to an individual's household or business; but the companies were permitted to provide the final hook-up from local base stations to individual premises by wireless, using the distinctive CDMA technology, not by copper wire. They gave their subscribers a cordless, 'loaf of bread' phone which, in theory, was intended to be used within the confines of a home or office, but in practice gave customers mobility within the localities covered by Reliance or Tata transmission towers.[31] The effect was spectacular. By 2004, there were 48 million connections supplied in this way, and they already outnumbered the 41 million phones connected by genuine, copper-wire-connected landlines. A year later, the Wireless in Local Loop (Fixed) connections, as the cunning new option was called, had almost doubled to 86 million, while the number of conventional landlines remained at 41 million. (In addition, genuine mobile-phone numbers by 2006 reached 90 million).[32] The Cellular Operators Association of India (COAI), bitter rivals of the Reliance and Tata companies, bemoaned 'the sickness of the industry which is facing unequal and illegal competition' likely to cost honest operators 'around Rs 18,400 crores (about US $3.7 billion) in losses' over the twenty years of their licences.[33] The Cellular Operators correctly concluded that the Government of India had decided 'that nothing should be allowed to stand in the way of pursuing the objective of increasing tele-density in the country' and that the great Ambani and Tata operations were to be the beneficiaries.[34] Eventually, in 2003 the Government of India did what the Cellular Operators had deplored: created a Universal Access Service Licence (UASL) which legalised the infringing Tata and Reliance businesses on payment of additional fees.[35]

These contests were fought out in the tall buildings of Mumbai and the red sandstone offices of New Delhi. The huge advertising campaigns of Reliance, Hutch, Vodafone and Airtel made more and more people aware of mobile phones. And as awareness grew so did the search for customers, and for sales people capable of capturing customers, even in small towns and the countryside. (See Illus. 8).

The art of retail

How the phone found its way into more lives, and carried capitalist practices with it, was illustrated in the development of a Chennai-based business. According to the company history, D. Satish Babu (b. 1967) began his working life as a door-to-door salesman of vacuum cleaners. He got into mobile phones in 1997 when he began selling brand-name phones with post-paid connections door-to-door for a company called Skycell in Chennai. Phones and calls were expensive; brand-name phones were uncommon; and much of the trade was in the 'grey market'— phones imported illegally from China (Chapter 4). One estimate was that 'hardly one per cent of the mobile handsets were sold through regular retail'.[36]

Sensing customers' desire to be gently inducted into the dark arts of mobile telephony, Satish Babu opened his own store in Chennai in 2000 under the name UniverCell. Intended as a place that explained the value of reliable brands, it aimed to provide customers with a comfortable environment to learn about phones and, more important, to try them out.[37] 'He wanted to provide the ambience', a company vice-president explained. 'He wanted to provide touch and feel, and he wanted to provide original mobiles with warranty'.[38] In 2007, UniverCell had grown to eight stores. Between 2007 and 2009, Indian phone connections doubled from 206 million to 430 million. UniverCell helped to lead this expansion. By 2010, it had 300 stores in south India and outlets in Mumbai and New Delhi; it claimed an annual turnover of Rs 600 crores

(US $120 million), monthly sales of 200,000 phones and a sales staff of 1,500.[39]

The mobile phone trained people in practices of international consumerism. UniverCell's first slogan had been, 'Where you buy matters', an attempt to reassure sceptical customers of the reliability that they could expect from a spacious, sparkling showroom and polite staff who taught people how to use branded phones that were sold with receipts and warranties. As the company expanded, it took this shopping experience—similar to what one might expect in Sydney or San Francisco—to 170 towns across south India. The slogan also changed to 'the mobile expert'. The assumption was no longer that people needed assurance that the phone would work—or be repaired if it did not—but rather that here was a place where one model could be compared with another and all the latest features could be tried and explained by trained staff.

The need for trained workers led to the diffusion of basic familiarity with mobile phones to millions of people. More confident customers became interested in more sophisticated technology. Sales people had to be prepared. An executive of UniverCell explained:

> There is a rigorous training program that goes on which is done partly by the manufacturers like Nokia, Sony, Samsung. They come and train our people. Second is [that] our own people train, get trained and they go about training the people in the showrooms ... We have 380 touch points. Now, you would call [those sales people to] the hub and ... give the training. And we also use video conferencing in a big way to train our people. ... Once in a month, there is a video conference happening where you list ... the focus models. What are the price ranges to focus on? Why are we focusing on certain models? What is our position of supply of stocks? For example, in a price segment, if you want a touch phone with a megapixel camera of, let us say, Samsung and that is not available, what is the alternate phone the staff can focus on? All this is done through rigorous training. And it is done frequently so there is a drill down ... to the last person. First, the [state] heads are trained, then the zonal managers, then the area managers, then the showroom managers, then the front end staff and so on. So this process is a well oiled machinery that keeps training and churning.[40]

Such systematic planning resonated well with a key element of global capitalism—standardisation.[41] What had worked for Henry

Ford and the American automobile industry was relevant to India's cell-phone growth.[42] Standardisation enabled global integration, and, like standardised components, which were easier to replace, so too the standardisation of skills made trained staff interchangeable. The mobile phone industry, led by global corporations, introduced new work practices that were shared across the industry and increasingly took root at the local level (Chapter 4). By 2010, UniverCell's stores covered most south Indian towns with populations of 200,000 people or more. The next level was to take the showroom and the customer experience to 'Tier 4' towns with populations of 100,000 to 200,000.

The difference in the treatment of customers was one of the striking features of the expansion of mobile telephony. Economy, comfort and education were all part of the mix. Curious citizens increasingly had to choose what qualities they valued and to weigh up the advantages of becoming a practising global consumer. In south India from 2000, Satish Babu's UniverCell offered customers friendly instruction and the chance to compare phone models. In north India, no chain quite like UniverCell emerged in the first decade after mobile telephony burgeoned. Instead, the big manufacturers—Nokia, Samsung, Sony-Ericsson and various makes of 'Indian brands'—either set up their own shops or induced existing outlets to add an agency for mobile handsets to their existing business. A few examples illustrate the process.

In November 2008, a young MBA from Allahabad University, backed by his family, paid a security of Rs 350,000 to Nokia to acquire an agency for their phones in a neighbourhood shopping centre in Allahabad in Uttar Pradesh. He opened a small store from which he sold Nokia phones and acted as an agent for Bharti Airtel. He combined the cell phone business with an agency for motor scooters. People who wanted one item, he said, were often interested in the other. Nokia widely marketed the virtues of their phones and paid him interest on his security deposit. There were other Nokia dealers in Allahabad, and there were many rival providers of mobile-phone cards and plans, but Airtel was the largest. In March 2009, after being open for four months, business was good. The shop looked reliable and reassuring, and the shelves were well-stocked with shiny boxes and fancy new phones. Nokia representatives visited regularly, and Nokia supplied additional stock as phones were sold. The young manager-owner said he sold about two dozen phones a week. Perhaps typifying the way many young people took up the technology, he said he had learned technical aspects of cell phones, not from industry training courses or tech-school study but from articles on the Web. In one respect, his was an 'old' business: his family were Guptas, traditionally associated with commerce. The young manager-owner said that he could have had a place in the family business, but it was better to work for oneself. He had chosen to sell phones and scooters because he perceived these as products whose popularity was likely to grow.

The cell-phone industry opened up new occupations and drew millions of people into unprecedented connections with corporations and governments. A second Nokia dealer in Allahabad in 2009 hinted at such possibilities. 'Rama Telecom and Photostat' had begun in 1998 as a business-supplies centre offering stationery, photocopying and express mail services. Adding cell phones responded to the interests of customers. At Rama Telecom and Photostat, Nokia was the phone on sale, but the service provider was Bharat Sanchar Nigam Ltd (BSNL), the government company. The business was owned by a Yadav, and a young Muslim dealt with customers at the front counter.[43] Categorised as OBCs (Other Backward Classes), Yadavs in the past were not regarded as people of commerce. Similarly, Muslims in north India since the Partition had been disproportionately artisans, peasants and workers. The Yadav-Muslim combination in the shop provided a cameo of new possibilities.

For customers, the new showrooms and mobile phone shops were places where they became familiar with the technical and social practices of a consumer-oriented world. By the first decade of the twenty-first century, even rural people were increasingly exposed to these practices. UniverCell executives explained that during festival times such as Diwali, country folk came to town and sometimes included the showroom experience as a feature of their holiday outing.[44] These interactions introduced people to new ways of life and purchasing practices. From Chennai to Banaras, sales people developed appropriately local ways of engaging with those who wished to buy a phone. And for many, a phone had become 'the first white good [major appliance] that a consumer buys, even before a bicycle'.[45]

Pivotal in assisting local entrepreneurs to seduce India's new consumers into the wonderland of mobile telephony were the sales promoters—spreading the word and explaining how to use the new device. This legion of missionaries propagated the message that mobile phones could and should be used by everyone, not only the privileged classes. Advertising campaigns celebrated lifestyles associated with mobile phones; but it was sales people down to the level of small shopkeepers who explained how a phone could be used in daily lives and suggested even to poor people that they too could be 'valued customers', entitled to a mobile-phone lifestyle. Small shopkeepers, and sales advisors employed by mobile phone companies, were trained in the 'art of retail' to educate first-time users about the world of mobile products and services.

The story of Ravi Varma, owner of a successful mobile phone shop near Banaras's busy Lanka crossing, captures the evolution of the industry from post-paid to pre-paid services.[46] His entry into the mobile phone business in 2001–2 was through Reliance mobiles as it tried to penetrate local markets. Ravi exemplified Reliance's vision of large markets in small towns, but it was a vision that remained largely unrealised until the arrival of pre-paid cards. During the early days, handsets were very expensive (about Rs 21,000 each), and the post-paid billing system was costly and unreliable. Even the sweeteners offered by Reliance to attract new customers, such as free talk-time as part of the post-paid bundle, were insufficient. Customers rarely made additional calls, fearing the unpredictably high cost of post-paid calls. But the real problem was more practical. 'Reliance', Ravi said, 'found it almost impossible to chase up their post-paid clients for their overdue payments'. Reliance managed to retain its hard-won post-paid clients, and once pre-paid schemes were introduced in 2003, the company shifted strategy. Ravi explained:

> The trick of Reliance was to turn all the post-paid connections to pre-paid, because they [Reliance] realised that people will not pay the money on the post-paid bills, so in order not to lose clients they kept their connections and converted these people to pre-paid.

Converting pre-paid was not simply a pragmatic move. For many consumers this was a transformative event. And it enabled people like Ravi to turn their fledgling ventures into profit-making businesses. Once the pre-paid mobile system began, people felt liberated, Ravi said; they could easily choose, and change, service providers and networks. They also realised that they could readily monitor their own call costs. At the same time, the prices of handsets were dropping, aided by the increased flow of Chinese mobiles and competition amongst service providers. For Ravi as a sales agent, abandoning his Reliance affiliation in favour of Samsung was a smart move; he soon became a 'privileged' Samsung outlet and adopted a new model to sell phones:

> I only had to invest Rs 50,000 and they gave everything—so I made my shop nice with air-conditioning. Also what was really good is that you only need to sell 60 per cent Samsung handsets and the rest whatever you want. But in practice this is not a problem as long as you achieve the target given by the company—which is about 250/300 sets per-month, and I sell around 1,200 sets in total per-month.[47]

Ravi's shop was spacious, with 12 glass display-cases sporting over 100 handsets, ranging from international brands, (e.g., Samsung, Sony, LG) to Indian ones, such as Fly, Micromax and Karbonn. (See Illus. 10). There were simple handsets as well as the top-of-the-range. The walls were decorated with posters of Bollywood stars advertising networks, brand-name handsets and various talk-time schemes. But according to Ravi, it was not about advertising; it was about trust:

> If I want to sell I can do it because we are the first point of contact with the customer and the relationship between me and customer is local. If I recommended a handset to clients, they will take [it] because if anything happens they will come back to me.

Ravi's clientele were mostly from the Lanka locality; many were students from the nearby Banaras Hindu University. Unlike other smaller mobile businesses that sold mainly mobile accessories, SIM cards and recharge coupons, Ravi's shop primarily dealt with handsets. The success of the shop rested on Ravi's experience and management skills, but equally important were the impeccably dressed and well-mannered sales promoters who arrived at the shop every morning at 8 am to begin their day. (See Illus. 9).

During the day the salesmen in Ravi's shop were behind the glass displays; three of the men worked for Samsung, the others for LG and Sony Ericsson. (See Illus. 10). These men were trained in the art of mobile phone retail. Their skills boosted the drawing power of the shop, as they sought to engage enthusiastic yet cautious consumers. Their nimble fingers easily manoeuvred between the different functions of the handsets; and their professional and polite manner added an air of sophistication and dignity to their interaction with potential buyers. Much of this involved educating people about the various functions and the latest applications of the mobile phones, the advantages of warranties, service centres, and of course, the respective brand-names. After all, they were trained and paid by the biggest multi-national companies in the field—Samsung, LG and Sony.

For many sales promoters in Ravi's shop, a routine day was twelve hours with a half-hour lunch break. The average monthly base-salary of these promoters was Rs 8,000, but, as they explained, this sum was usually topped up by 'incentives' offered by their respective companies, once they achieved their sales target for the month. For example, a Samsung promoter called Ashish noted that he once made Rs 15,000 in a month, having sold over 140 handsets, a figure confirmed by Ravi who kept close watch on all transactions in his shop.

Anil Kumar, one of the most experienced sales promoters in Ravi's shop, told how he began working in the cell-phone business. Anil was not a local, and like many of his colleagues, he lived in a rented flat and visited his hometown only every few months. Anil was originally from Gorakhpur (a city north of Banaras near the Nepalese border) where he trained as a Motorola salesman in 2007. The monthly starting salary at the time was Rs 3,500 (US $88) plus bonuses—a significant sum for a young student from a provincial town. After he graduated with a BComm degree from Gorakhpur University, he moved to Sony Ericsson where he saw more opportunities. Sony did not employ sales promoters in smaller towns such as Gorakhpur (population about 700,000), so he found a job in Banaras. He had been an employee of Sony Ericsson ever since and had worked at three different shops with Sony display cabinets. His base salary steadily increased to Rs 8,000 by 2011. He listed the benefits he found in working for a company like Sony Ericsson. Sony ensured sales promoters were kept up to date with the most recent handsets released to the market. The company arranged monthly refresher courses, usually in one of the top hotels in Lucknow, the state capital, and where he and his colleagues met with senior sales executives to exchange ideas and learn about new models.

Some sales promoters worked throughout the year, especially those based in larger outlets and showrooms. Others were employed as casuals, sent by the companies to boost sales during festival seasons, especially during the months leading to Diwali in October-November. This was common practice in smaller shops, where a big brand such as Nokia maintained a cabinet of their products but did not have a regular sales representative. This was the case with Shri Ram Communication—the neighbourhood mobile shop at Assi Ghat in Banaras. Unlike Ravi's air-conditioned shop, with its clean floors and smart-looking sale promoters, Shri Ram Communication was small (5 x 4 metres) and darkly lit, located near the busy Assi Ghat crossing. The shop walls were covered with handwritten posters promoting the latest pre-paid talk-time deals. Shri Ram largely catered to a local male clientele, and it was rare to see students or women in the shop. The majority of clients were from low-income families from the surrounding rural areas: 'regulars', comfortable entering the shop in their daily attire of a *lungi* (lower cloth) and *gamchcha* over their shoulder (a cotton towel for wiping sweat from the face) and speaking to the owner and his two workers in their own Bhojpuri language. The working day also followed local rhythms. Trading hours were flexible and imprecise; opening time hovered between 9 and 10 am, and Samir, the owner, and his two employees, could stay open until 10 pm if trade warranted.

Doron was introduced to Samir through a long-time friend, Deepak, a boatman. Deepak and his colleagues regularly visited Samir's shop to buy pre-paid credits on their way to the river or back home. Many of the boatmen, and other men in the neighbourhood, were semi-literate and required help in recharging their phones, something that Samir and his two employees were happy to do. Indeed, the majority of transactions involved selling small pre-paid talk-time units, mostly in the form of electronic-recharge (E-recharge) performed by the employee and where the client received an immediate SMS notification that the mobile had been credited. Some clients came to purchase SIM cards. Samir and his workers helped them to fill out the necessary forms and make photocopies of ID cards and of letters (usually a bill of some sort) verifying a person's address. Occasionally, someone would come in to buy one of the low to mid-range handsets that Samir kept in stock. The busiest months for selling handsets were those leading up to major festivals, especially Diwali, when people often purchase new products. During peak times Samir kept more than double the usual number of handsets. This was also when Nokia sent its sales force to various outlets to promote its products. While Samir kept various local brands in his shop, the main display case featured Nokia phones. (See Illus. 11).

In late September 2009, a garland of fresh flowers adorned the cabinet display every day. A young salesman sat next to the display case and sprang up every time a client ventured towards it. This was Sumit, the Nokia sales promoter sent to work in Sri Ram Communication for the couple of months leading to Diwali. The soft-spoken young man had been working in the store for just over a month when Doron met him. Sumit's story of how he was recruited and trained for the job, and of the pressures a sales promoter faced, was fairly typical of those of other sales promoters we met.

Sumit Churasia was only 20 in 2009, born and bred in the neighbourhood not far from Samir's shop. He recalled his first handset purchase, a Nokia 1600, back in 2006. Since then, he said, he had six different mobiles, including Samsung, LG and Reliance. Two of his older sets he sold to friends, and the rest were kept at home for family use. His father was a manual labourer, and like the majority of people in Banaras, their home had never had a landline telephone. Sumit's enthusiasm for mobile phones drove him to go for an interview as a sales promoter for Nokia, which took place at one of Nokia's major outlets in Banaras. He and 20 other people, only two of whom were women, were interviewed for about fifteen minutes each. Sumit was asked to give a short sales-pitch describing the features of his own mobile phone. While Sumit felt he had done well in the interview, it took Nokia more than a month to contact him. What followed was an awkward exchange between Sumit and the Nokia representative, as Sumit recalled:

> One day I saw a missed call on my mobile which I did not recognize. I returned the call asking whom I was speaking to, but the person on the other end of the line replied with the same question (*kaun bol raha hei?*). I insisted he disclose his name first because he gave me a missed call. But he refused and we began to argue, and I abused him for refusing to say his name. Eventually he said he was from Nokia, to which I replied, 'Yes, sir, please tell me' [in English]. His answer was: you are rejected (*tum reject ho gaye ho*), so I said, 'Why do you ring to tell me I am rejected a month after the interview?' He replied: 'An hour ago I

was going to tell you that you were selected, but now you are rejected'. So I tried to explain myself and apologized for my rude behaviour, but he just hung up the phone (*'unhone phone kaat diya'*). I then tried to call him back and explain myself several times but he rejected all my calls.

Yet Sumit was determined, and devised a strategy to contact the Nokia representative to explain himself. He tried calling back that night:

> In the evening I called that [the Nokia person's phone] phone from a different SIM, but he did not answer. He then called back later, and asked, 'who is this speaking?' I replied saying, 'you called me; who are you speaking', this time he began abusing me for giving a missed call and refusing to disclose my name. I then revealed my name asking him to recall our conversation in the morning and to try and understand my rude response. So eventually he said okay, I'll give you another chance, come tomorrow to the hotel Gautam Grand at 9 am to begin training.

There are a number of things to note about Sumit's account of his missed-call encounter. The first is to emphasise the hunger for work among many young men across UP, who remain unemployed despite their educational qualifications.[48] Many are desperate for work, whether they are people like Anil who relocated from Gorakhpur to Banaras after completing studies, or Sumit whose desperation was expressed in his persistence. These men were the worker-bees of global capitalism: both its enthusiastic consumers and its youthful, qualified promoters.

The account also reveals the somewhat disorganised and unpredictable recruitment process. This contrasted with the impressively organised Nokia training course that Sumit attended the following day. Over eight days Sumit and about thirty others enjoyed the luxurious air-conditioned setting of a hotel conference room where they received breakfast, lunch and tea from 9 am to 6 pm. The training involved learning about Nokia's main handset models from the cheapest to the most sophisticated and expensive. The latter part of the course focused on customer relations and methods of engaging clients and evaluating their needs:

> We were trained about how to behave with customers, what kind of language we should use and to avoid doing what many shopkeepers do, that is, to try and sell the most expensive phone to the consumer. Nokia follows a different method, our motto—'We don't sell mobiles—we sell solutions'.

MISSIONARIES OF THE MOBILE

Led by two instructors from Kolkata, the course used a range of teaching methods from power-point presentations to role-playing. The last day featured an exam and 'a role-play test'. A few days later, Sumit received a call notifying him of the outlet he was assigned to: a shop with a Nokia-registered display cabinet. This was what he called 'on the job training', followed by a 'refresher' conducted by Nokia a week later where the new sales promoters met their colleagues, team leaders and instructors to discuss their experiences.

Unlike the sales promoters working in the larger shops, Sumit worked only during peak times, primarily for the two months prior to Diwali and a few weeks after. According to Sumit, only those who excelled in the job got a permanent posting at one of the major Nokia outlets. When Doron met Sumit at Samir's small shop in mid-2009, Sumit had only been several months in the job. He kept a meticulous account in a small notebook of all the transactions in the store involving handset sales, including the models he sold, as well as the sales of competing brands, such as Samsung or Indian brands. (See Illus. 12). This information, he explained, was communicated twice a week to his Team Leader to account for his sales and to request new stock. 'We write in our logbook who bought a Nokia set from us and which model and also those who bought other brands and why, so that we tell this information to our team leader'. Thus, using its network of sales promoters, Nokia gained information about purchases and the changing needs of consumers. Such data could be used to respond appropriately, whether in pricing or features. An example was Nokia's eventual release of a handset capable of taking two SIM cards, something the company resisted, but developed in response to insistent consumers.

Sumit viewed his role as vital not only for communicating clients' demands to the company but as the foundation for the whole sales enterprise. 'We are the most important in the chain', he said, 'because if we [sales promoters] don't meet our target the Team Leader is also in trouble and Nokia too'. It is not surprising that Nokia and other corporations put considerable resources into training promoters and equipping them with skills to handle phone technology, relations with customers and record-keeping and data management. Sumit did not come from a 'traditional' business family whose members may have been keeping books and registers for generations. His father worked as a manual labourer for a local cloth company. Sumit was the first in his family to be introduced to bureaucratic capitalism and the requirement to report to unrelated, often unknown superiors. Though the modern managerial environment was new to him, he adapted to it enthusiastically: 'We are the most important in the chain!'

Medium-sized mobile shops such as Ravi Varma's Samsung outlet or the smaller Shri Ram Communication shop at Assi Ghat did not rely solely on formal networks to source their goods. This was especially the case in the business of mobile accessories, such as SIM cards, recharge coupons and accessories, which drew supplies from a wide network. Extended in various directions, the networks that were tapped often depended on the nature of the shop. In their investigation of mobile-phone shops in a Mumbai slum, Rangaswamy and Nair argued that such 'hybrid networks' were a typical feature of small mobile-phone enterprises, where shopkeepers

> forge relations with procurement channels like grey markets, mediate between mobile phone companies (even multinational corporations) and the consumer, expand business loops while renewing existing ones, and encourage apprenticeships for relevant repairing skills.[49]

The next chapter examines the repair industry in more detail, with its 'informal' networks and practices.

As the enthusiastic Sumit suggested, the missionaries of the mobile were connected in a chain. At one end were urbane advertising executives struggling to transform international marketing concepts about consumers and communications into messages that India's huge diversity would respond to. In towns and cities, an experienced marketer observed, 'there is so much clutter; they are getting about a hundred messages a day'; but 'in rural India the challenge is just getting to them'.[50] Once mobile telephony became rewarding for business in 2003, the big enterprises that sold phones and provided cell-phone services drew on the long-standing connections of India's Fast Moving Consumer Goods industry. If profit lay in mass sales of phones and services, FMCG networks provided a path: 'FMCG has been using the retailer as an avenue of communication for a very long time'.[51] The education of retailers, the widespread arousal of interest and the creation of sales and support networks brought mobile phones within

MISSIONARIES OF THE MOBILE

the consciousness and the reach of more Indians than any previous consumer item. At the 'bottom of the pyramid', the cell phone industry offered both sellers and users more ways—and sometimes *new* ways—to earn their living.

4

MECHANICS OF THE MOBILE

> *The automobile itself became a major business, spawning dealerships, service stations, [and] repair shops.*
>
> Claude S. Fischer, *America Calling*

When the Supreme Court of India revoked licences for 2G Radio Frequency spectrum awarded in a controversial allocation in 2008, one of the affected companies instantly sought sympathy by appealing to 'the authorities ... to ensure that our ... 17,500 workforce and 22,000 partners are not unjustly affected'.[1] Mobile phones created jobs in manufacturing, distribution, maintenance and repair. In 2007, 2.5 million people were estimated to be employed in the 'telecom services sector', and this figure did not include the most innovative mobile mechanics of all—those doing repairs and second-hand sales in small shops in every town in the country.[2]

This cascade of new occupations, and the training and skills that went with it, resembled to some extent the expansion of the automobile industry in industrial economies after the First World War. In the US, the popularity of the automobile created 'a system of dealers who not only offered sales outlets ... but also service facilities'. A 'used car market blossomed' from the 1920s.[3] In every hamlet, petrol (gas) pumps appeared, and people were employed to pump gas and provide 'service'. 'Service stations' became a feature of American life, as did the mechanics who were part of them. As late as 2012, when the American car industry was thought to be in alarming decline, it was estimated to employ 3.5 million people, 800,000 in repairs and maintenance.[4]

If cars in the US bred like rabbits, mobile phones in India gushed like water from a burst dam—rapidly and into every nook and corner. A dam *had* burst: the dam that limited what people in villages, and even towns and cities, were able to know and how they were able to communicate. Those barriers, as we have seen, were structural (poor roads, limited radio and television, relatively few newspapers) and ideological (some people were *entitled* to know more than others and some people were positively forbidden from knowing some things). The popularity of the mobile phone generated a need for workers who could make, repair and sell them. The two or three million people estimated to be involved in provision of telecommunications were simply those in formal occupations identifiable for government statistical purposes. They included thousands of (mostly) young rural women who found jobs in the manufacturing of phones. But many more people were involved in the informal repair and resale industries that flourished in every town. They were the 'mechanics of the mobile'.

And another kind of 'mechanics' was also at work: the process by which people learned about their phones—how to buy, sell, fix and use. As we saw in Chapter 3, great corporations and small shopkeepers selling Fast Moving Consumer Goods became part of the education system for cell phones. But many people found neighbourhood mobile mechanics more to their taste and budget.

People

FACTORY WORKERS

At Sriperumbudur where Rajiv Gandhi was assassinated, Nokia had its largest factory in the world in 2010. About 12,000 people were employed on a vast, carefully controlled floor, located in a Special Economic Zone (SEZ) of 85 hectares. The factory ran 24 hours a day with three eight-hour shifts six days a week. Most of the employees were young women, and the story told by Nokia celebrated the effects of mobile-phone manufacture on society and economics in Tamil Nadu.

MECHANICS OF THE MOBILE

'Seventy per cent of our employees are women', said Sabyasachi Patra, head of government relations for Nokia in the Chennai area in 2010. 'The reason is when we take these people, we do [a] dexterity test and hand-eye coordination test. It's the woman who gets through'.[5] To qualify to take the test, applicants had to have twelve years of schooling and passed final exams with a mark of at least 60 per cent. '90 per cent of our employees are from economically poor background. Even one-meal-a-day kind of families. You'd be surprised at how poor some of them are'. On a starting wage of about Rs 5,000 (US $100) a month, factory work brought training, free transportation, meals and medical care. For some, it could also mean the possibility of further study. 'After saving for one and a half years or so, they are able to save Rs 50,000 or 60,000, something like that. They can go for higher studies, and some of them are', Patra continued. In some cases, Nokia paid the costs in return for bonding the employee to the company for two or three years after completion.

The workers were not, according to the company account, a classic urban proletariat. They were recruited largely from villages up to 60 kilometres west of Sriperumbudur in rural Tamil Nadu, not from the great city of Chennai to the east. They were mostly young women, were relatively well educated and were living with their families. 'We spend a huge amount of money', Patra said, to pick up and deliver people from their villages over such a wide area. Most of the women were aged between 18 and 22, and the factory had a fairly steady turnover because once married, 'families don't like it if the wife or daughter-in-law goes to the factory working night'.

Nevertheless, the company story emphasised the social change that efficient, dignified factory employment brought about. Most of the workers were in 'first-time employment', Patra said. On one occasion, a group of young workers had been called to meet a gathering of foreign guests. One young woman said, 'Yes, money is important, but it's just not money'. What else was it?

> It's very important for us that I get Rs 5,000. My father gets Rs 2,000 a month; I get Rs 5,000 a month. What I like is, I now go back home and tell my brother what to study, tell my father what to do, tell my neighbours what to do.

Patra added his conclusion: 'So this girl, in her locality, is seen as a leader'.

There was, however, a different view of life in a mobile-phone factory. *Phony Equality*, a report on workers' conditions in the four factories in the Nokia SEZ at Sriperumbudur, was published in 2011. Conducted by three NGOs, two from Europe and one from India, the report was based on interviews with a hundred workers and visits to the factories by researchers from two of the NGOs.[6] Contrary to the Nokia contention, the report asserted that 'most of the workers lived in rented rooms', not at home in their villages. 'The vast majority ... were migrants ... and ... had to live in temporary accommodation'.[7] Presenting the condition of workers as underpaid and precarious, the report carried pictures of 'a room and kitchen shared by twenty-five working women for Nokia and Foxconn'.[8] It confirmed the Nokia assertion that 'the mobile phone manufacturing workforce is the first generation of industrial workers in this region with parents who have been agricultural labourers throughout their lives' and held that this made the workers especially vulnerable to exploitation.[9] 'For poor, rural youth, a factory job with an identity card and uniform offers a degree of social status'.[10] Most of the workers were trainees or on contracts, hired by employment agents and not directly by Nokia or by the other three companies operating in the SEZ.[11] Labour laws were 'routinely flouted',[12] and contracts and trainee systems were said to 'trap workers in an unfair and exploitative position'.[13] 'Most of the surveyed workers' believed that their experience in the mobile-phone industry was "a waste"' and could not see how it would advance their 'career prospects'.[14] The companies had accepted some unionisation after strikes in 2009 and 2010.[15] Allies of the protesting workers put a video of demonstrations on YouTube—shot on workers' mobile phones.[16]

The Nokia factory, however strained and contentious its labour relations, could turn out 300,000 phones a day. Elsewhere in India, the old government telecom factories in Bangalore represented a different model of production and labour relations. The Bangalore factory of Indian Telecom Industries (ITI) was the 'first state-run enterprise' in independent India, started in 1948. The travails of ITI became the subject of a 680-page book, among whose many details was the fact that in ten years, the factories had manufactured 9 million phones, a total that the Nokia factory could have equalled in a month.[17] The government factories suffered from overstaffing, poor discipline, outdated products and promotion solely by seniority. 'A role reversal ... occurred where it was in the supervisor's self-interest to maintain cordial relations with his subordinates ... Misdemeanours ... generally went unreported ...'.[18] The private-enterprise mobile-phone industry of the twenty-first century forced some changes in the government factories.

The ITI factory located in the small town of Raebareli in north India provided an illustration. Originally established in 1973 for manufacturing landline equipment and infrastructure, the ITI factory in Raebareli began to change its work practices. In 2005, it began to try to cater for the demands of the thriving mobile phone industry and particularly for the state-owned telecommunication company, BSNL. The factory, which employed more than 3,700 people in 2011, had been primarily making and installing mobile phone equipment for transmission towers.[19] From 2005 it produced and assembled more than 2,000 towers a year for urban and rural markets. Its hi-tech electronics division, trained by the French telecom giant Alcatel, staffed the sophisticated machinery for assembling Base Trans-Receiver Station (BTS) equipment. The factory had incorporated some of the rituals of modern factory practice into its daily routine, with 24-hour shifts, and the token requirements to use health and safety equipment, which the staff still managed to ignore. The Raebareli factory remained almost exclusively staffed by middle-aged men, most of whom sat gazing at the machines, waiting for their shifts to end.

The contrast between the government factories of ITI and the Nokia factory of the Special Economic Zone highlighted the different experience that cell phone manufacturing meant for its workers. In the ITI factories, the workers were largely men of mature years, highly unionised and treating the job as lifetime employment, regardless of whether production targets were met. In the Nokia and associated factories, workers were mostly young women, often with little job security, tenuous links to trade unions and a keen awareness of the need to keep the production line moving.

There are unattractive features of both pictures: sloth and slovenliness in one; exploitation in the other. But the young rural women exposed to factory life illustrated another break with the past that the mobile phone provoked. Factory conditions may have been harsh, but they threw people together, just as Marx and

Engels had seen in Victorian England. Out of such random but large-scale shoulder-rubbing, new consciousness would emerge, though not necessarily the militant class consciousness that Marxists predicted. The mobile phone was tiny. Unlike the factories of the nineteenth and twentieth centuries, manufacturing mobile phones required dexterity, not physical strength. Women did it better than men, and young women, the big companies concluded, were easier to control than men. In leading thousands of young women into wage-earning, strictly disciplined, card-carrying factory life, the mobile phone industry may have taken on the role of the munitions factories of the World Wars—bringing women into the workforce in ways that would change social practices. Though the numbers were small relative to the population of India (tens of thousands, not tens of millions), their employment was part of a cascade of different and often new jobs that accompanied mobile telephony.[20]

TOWER WALAS

Tower building and maintenance was another way that the cellphone industry infiltrated the lives of large numbers of people, many previously little affected by the perils and incentives of modern states and capitalism. In 2012, 400,000 mobile phone towers were estimated to dot India, each one of them requiring landlords to offer them space, managers, lawyers and clerks to tie up the deals with the landlords, engineers and technicians to care for the computers and air conditioning equipment, delivery people to transport the large quantities of diesel fuel that powered the back-up motors, skilled workers to maintain the towers against wind and corrosion and security people to guard against theft and vandalism.

Mobile phone towers were an industry in themselves, and companies were set up simply to erect and maintain towers. The largest was Indus Towers which operated 110,000 towers in 2012.[21] Established in 2007, it was a consortium of three of the largest service providers—Airtel, Vodafone and Idea, which together claimed more than 45 per cent of Indian mobile phone users. Indus Towers rented space on its structures to its founding companies—and any other company that wanted to pay the rent—and claimed 200,000 tenancies on its sites in 2012.[22] The structure of the

MECHANICS OF THE MOBILE

company, and the contracts that bound it together and under which it rented out its facilities, belonged to corporate India, far away in the towers of Mumbai and Delhi. But the transactions that put towers into paddy fields and onto people's rooftops took the industry into tens of thousands of homes and villages.

The state of Kerala constituted just over 1 per cent of India's land area, but to give its hilly terrain almost complete phone coverage required 12,000 towers in 2011. The leading tower builder was India Telecom Infra Ltd (ITIL), set up in 2007 as a joint venture between the TVS Group, a south India conglomerate that began as a bus company before the First World War, and Infrastructure Leasing and Financial Services Limited. Bobby Sebastian, a veteran of the telecom industry, headed ITIL's Kerala operations in 2010 and explained what a phone tower company had to do to be profitable.[23]

Service providers used Global Positioning Technology to work out the location in which they needed to place towers to provide coverage for a specific area or 'cell'. Radio Frequency in the bands used by mobile phones needs a clear path from one tower to another, hence in hilly country more towers are needed to prevent dropouts and ensure good service. Once telecom companies identified areas where they needed towers, they passed the locales to tower-building companies who sent acquisition agents to identify sites where equipment could be installed or towers could be erected. Sebastian's company would offer service providers three possible sites. The provider selected the one that best suited them, and the tower company approached the owner of the land to complete a rental agreement. The tower company then applied to the local government for a permit to build the tower, called in contractors to build it and got a completion certificate after inspection of the tower.

When the structure was complete, service-providers installed their electronic equipment. The more providers Sebastian's company (ITIL) could persuade to rent space on the same tower, the better for ITIL's business. It cost little more to service a tower supporting seven providers than a tower with only one. ITIL paid the electricity and diesel bills and was reimbursed by the service providers. ITIL also dealt with local governments and with thousands of individual owners on whose property the towers sat. The tower company remained responsible for the diesel motor that

95

ran the back-up generator for the air conditioning, as well as for the electrical connection to the power grid, the maintenance of the tower and the security of the tiny area around the tower and its air-conditioned shed. The shed presented temptations, not so much for its air conditioner or its electronics but for its valuable and readily cashable diesel fuel.

Diesel posed problems. Diesel-powered back-up generators were essential for every tower because of the irregular supply of electricity and the dependence of cell-phone technology on constant electricity. The electronic equipment in each tower required air-conditioned comfort—not too hot, not too cold. Diesel was polluting, and diesel was costly. BSNL in Chennai alone spent Rs 3.5 crores (US $700,000) on diesel before it paid its electricity bills.[24] As demand for mobile phones and better services grew, so did the number of towers and the need for diesel.

Towers came up in breathtaking places: deep in rubber estates, in church and temple compounds, on top of houses, in gardens and open fields and on ruined hilltop forts. The latter were especially useful since they had been built hundreds of years earlier for the same reason that made them valuable in the twenty-first century: a good view across a wide area. For property owners who signed a 15-year lease, hosting a tower brought a monthly rent of about Rs 6,500 (US $150) in rural Kerala in 2010 and up to Rs 20,000 (US $470) in a city like Kochi. These were handsome sums when a new worker in the Nokia factory outside Chennai drew Rs 5,000 for eight-hour shifts six days a week. In cities like Lucknow in north India, advertisements in Hindi daily newspapers offered up to Rs 30,000 (US $600) a month for tower sites in areas crucial for coverage. (See Illus. 14).[25]

The towers were to telecommunications what gas stations had been to the automobile: a physical presence that drew together threads connecting large numbers of people and interests. If India had 400,000 towers in 2012, agreements had been executed with hundreds of thousands of landowners. All those agreements, covering towers of varying heights, foundations and wind resistances, should have been scrutinised by the clerks and inspectors of local governments. Once constructed, towers needed constant checking and maintenance—one person permanently employed for every four or five towers, by Sebastian's calculation. At that rate, close to 100,000 people in India earned a living through tower maintenance alone.

MECHANICS OF THE MOBILE

MISTRIIS[26]

The manufacture of phones, and the construction and maintenance of towers, were carried out by capitalist enterprises that imposed the formalities of modern business and bureaucracy. But the revolutionary character of mobile phones lay in their sheer numbers—the fact that they were cheap enough for hundreds of millions of people to own, and phones failed every day. When they failed, people fretted and quickly sought repairs. Fixing phones, like skinning cats, could be done in different ways. One way was to go to a local hole-in-the-wall shop where a local man—invariably a man—had set up a cottage industry as a mobile phone repairer. The need for repairs required a distressed phone owner to make choices.

Lucknow, the capital of Uttar Pradesh, bustles, and Hazratganj, its centre, pounds like a heart. Huge billboards, snail-paced traffic, modern shopping complexes and multiplex cinemas sit beside historic buildings and time-tested restaurants serving Mughlai food. In June 2010, it was over 40 degrees when Doron met Saif Siddiqui and his brother-in-law, Salim. Their tiny 1 x 3-metre shop was one of three in a row, all occupied by men repairing mobiles. Doron visited them a number of times, leaning on the counter that separated them from passersby in the street. Saif and Salim were Muslims in their mid-twenties. Saif opened the shop in 2005 after a friend suggested that it would be a more profitable business than the jeans-and-sunglasses venture they operated at the time. Soon after, he explained, a few more repair shops opened next to his. Saif left school at about fifteen after completing tenth standard. He gained his mobile repair skills from a friend, rather than in a training institute. According to Saif, it took him three years to become competent in fixing almost all types of mobile phones, including the latest 'smart phone' models. In 2010, his road-side shop was full of handsets in different states of dysfunction; he bought them cheaply from a wholesaler. (See Illus. 15).

Saif and his brother-in-law were *mistriis* or artisans. Throughout the day they engaged in 'grooming', fixing and adjusting all types of handsets for people off the street. Transactions were fairly straightforward. Saif and his partner sat behind the counter and interacted with their customers who stood on the pavement, sometimes inspecting the devices on display in the very-mini showroom. The customers explained the fault and asked for a quick evaluation. The most common faults, Saif said, included water damage, connectivity problems and screen or recharging faults, often caused by moisture. The moisture problem was usually solved by prying open the phone and dipping the motherboard in a bucket of bad-smelling chemicals that Saif kept next to his bare feet. A toothbrush was used to clean and dry the board before reassembly. Charging about Rs 100 to fix the most common faults, Saif earned on average Rs 800–1,000 per day. When there were no customers hovering over the counter in anticipation of their device being fixed, Saif turned to the more challenging cases that required careful diagnosis and repair. Saif used a computer in a corner of the shop for software reformatting and downloading songs, wallpapers, video-clips and films. He charged about Rs 50 for one gigabyte of memory download.

Sometimes, when he was unable to repair a device, Saif consulted a more experienced friend who worked in a nearby mobile phone shop that sold various branded and Indian mobiles. If Saif could not fix the problem, clients were not asked to pay. Indeed, Saif might offer to purchase the handset if its parts were useful. Otherwise, clients might opt to try to repair it elsewhere.

Most of Saif's clientele came to him for handset repairs. Many customers belonged to Saif's Sunni community, confirming an observation of Nimmi Rangaswamy and Sumitra Nair that small businesses draws on community and kin networks.[27] According to Saif, his clients were mostly men because, even if the mobile was used by a woman, it was the men in the house who had the responsibility for maintaining and fixing household electronics. And the stalls of the streets were male-dominated public places.

Towns and cities in Uttar Pradesh were saturated with similar small businesses retailing mobile phones and associated products. 'Grey markets' harboured more specialised outlets, where one could find a string of shops and make-shift stalls that only repaired mobile phones, both their hardware and software.[28] In Banaras as in Lucknow, the *mistriis* in the grey markets were credited with being experts, and if the neighbourhood *mistrii* was not up to the task of fixing a handset, he would refer it to the 'higher authorities' located in the grey markets. In Banaras that meant the Daal Mandi market located in the older part of town.[29] The mobile phone section of Daal Mandi grew out of what was once a thriving electronics bazaar, dominated by pirated DVDs and CDs, televisions and other electronic goods. Muslims, who were the dominant group in the area, traded mostly in second-hand and non-branded phones, known as 'China Mobiles'. While most handsets, including branded ones, were made in China, the term 'China Mobile' referred to non-branded, flashy handsets. In mid-2009, these mobile phones were considerably cheaper than the branded devices but often claimed to have many of the same multimedia functions. Some even promised much more—such as a 30-day battery back-up, 8-megapixel cameras and TV reception. The stalls and shops in Daal Mandi offered a huge pool of components for replacing and cross-fertilizing mobile phones. For example, 'G five' handsets had batteries, screens and other components that fitted easily into Nokia or Samsung models, thus making factory-original components less necessary. (See Illus. 16).

China Mobiles were the first to feature two and three-SIM card handsets. This was a very attractive feature, especially for poorer consumers who often alternated between SIM cards to save money by taking advantage of varying rates offered by different service providers. At different times of the day or week one provider or another would often offer cheap rates to attract customers; for customers, it was attractive to be able to switch operators by tapping quickly from one SIM card to another. Nokia officials admitted ruefully that Nokia had been slow to react to India's love affair with dual-SIM phones and to start manufacturing its own models.[30] China Mobiles, unlike their legitimate relatives, were sold with no guarantee, warranty or bill of purchase. Their price was much lower, not only because they were non-branded and of lower quality, but also because they were not connected to any service centres and the retailer did not pay Value Added Tax (VAT).

The service and repair provided in the grey-market economy was vital for a number of reasons. Because China Mobiles came without warranties, repair-walas and their stalls provided the only way of fixing a faulty cheap phone. But many branded mobiles that did not have warranties also ended up with the *mistriis*. Although branded mobiles such as Nokia, Samsung and LG were popular with poorer people if they could afford them, the services associated with these branded products often seemed inaccessible—too unfamiliar, corporate and forbidding. Moreover, once a mobile was tampered with, not only was the warranty void, but it was instantly rendered an 'untouchable item' by brand-name service providers who refused to have anything to do with it.

In Daal Mandi in Banaras, the cost and types of a handset varied slightly across the thirty-odd stalls. Business thrived in the alleys, not only among the retailers but also in the repair businesses located in the narrow lanes. The small entrepreneurs procured their supplies of mobile phones from larger markets in Delhi or Mumbai. For example, Arif and his brother who operated a shop in Daal Mandi explained that they got their mobile phones from Ghaffar Market in Delhi. Their shop had around forty Chinese-made mobile phones in all ranges and prices. Arif sold about ten sets a day at an average cost of Rs 1,500 (US $30). There was a constant flow of customers inspecting, comparing and negotiating. For these shoppers, this was an attractive and familiar way to shop and a place where the consumer wielded significant power.[31]

TRAINERS AND TRAINEES

How did *mistriis* of the mobile learn their skills? The automobile industry offered parallels. 'A [Model T] Ford', E. B. White wrote, 'was born naked as a baby, and a flourishing industry grew up out of correcting its rare deficiencies and combating its fascinating diseases'.[32] The difference, however, lay in the fact that many more people in India aspired to talk than ever aspired to drive. The demand was far greater for the mobile-phone equivalents of what White called 'the heaven sent mechanics who could really make the car talk'.[33] When their phones died, people looked desperately for those who could bring them back from the dead.

In Mumbai slums, Rangaswamy and Nair met store-owners who had trained in phone repair in formal courses at institutes charging US $300 for a four-month course. For people living in the Mumbai slums of Behram Baug, such sums were substantial, and for most of the trainees an additional period was required in which they engaged in 'self-training on the job' or supplemented 'their training through peer-learning'.[34]

Doron's inquiries in Lucknow, Banaras and New Delhi yielded similar findings. While many of the street-repairers were trained by their peers and gained proficiency on the job, the most competent had formal training in an institute. In some instances, a family invested in training one member at an institute so that he could pass on the knowledge to his community and relatives. (Doron did not encounter a woman *mistrii*.) But terms such as 'peer-learning' and 'formal training' needed explaining when used in connection with mobile phones.

The repair industry was highly competitive, and informants explained that they only shared their knowledge with close kin and friends. This had implications for the way they conducted their trade and improved their skills. According to one repairman in Banaras, when he sent mobiles that he was unable to repair to Daal Mandi, he marked the components with microscopic precision so that the repairers in Daal Mandi did not replace them with used parts. He explained that these more skilled repairers jealously guarded their knowledge, restricted access to their shops and barred others from viewing their work. The microscopic marking had another function: to increase his skills in identifying the source of a problem. He would mark several suspect components in the faulty device, and when the expert returned the repaired mobile, he would open it, not only to ensure no used parts had been inserted, but also to try to identify the fault by checking which of the marked components had been tinkered with. 'Peer learning' could be by stealth as well as by sharing.

There were varying degrees of formality in the training process.[35] Informal training could be gained by apprenticing in a shop or learning from peers, while the formal acquisition of skills was more systematic. When Doron asked a seventeen-year-old repairman in a small mobile-phone shop about his training, the young man referred Doron to his teacher located near the city centre of Banaras. The teacher's place was tucked in a small passageway next to a few photographic and mobile-phone shops. The teacher, Ajeet Singh, was a soft-spoken man in his early thirties who said he gained his skills by watching others and through reading repair manuals. His shop contained all sorts of mobile-phone accessories and parts, but Ajeet spent most of his day carrying out repairs, which he enjoyed because of the challenges they posed. For several years he had conducted courses for young men who wanted to learn repairing. According to Ajeet, mobile repairing was an especially productive option for young men who would otherwise be in limbo or relying on government employment schemes, such as MNREGA (Mahatma Gandhi National Rural Employment Guarantee Act), which offered a wage of only Rs 100 a day. Ajeet had trained more than sixty students; all had found jobs; and some had opened their own repair business. The course lasted about three months, for which he charged Rs 8,000 (US $160). The timing was flexible, but the student attended around three hours a day. The course began with a few weeks of instruction about hardware and identification and repair of faults, followed by study of software repair programs and techniques. After the course the student usually spent a few weeks with Ajeet to gain on-the-job experience.

Ajeet had a number of large books that he regularly ordered from Jaipur's G Tech publishing and that could also be purchased in Daal Mandi. These soft-cover volumes offered a range of mobile repair solutions from basic tool kits to highly detailed diagrams of mobile-phone circuits, parts and fault-finding techniques. Other books were specific to downloading and software solutions. In Hindi and English, they featured the latest cell-phone models, ranging from Blackberry, Nokia and Samsung to the unbranded China Mobiles.

The software program most commonly used to unlock and revive mobiles was the UFS3, specific to Nokia models. For most other devices, including Samsung, LG and Indian brands, such as Spice and Karbonn, as well as China Mobiles, the commonly used software was known as Spider Man. Ajeet added that one should be cautious when using software as one could easily alter the IMEI number.[36] This could be dangerous since mobile phones attracted both terrorists and security forces, and, in any case, tampering with a phone's unique identification number was illegal and could result in unwanted attention (Chapter 8).

The rapid, widespread uptake of mobile phones explained why phone repairers were in high demand. To put the mobile phone in perspective, the total production of the Model-T Ford, the product that made the automobile a mass-consumption item in North America, was 15 million cars between 1910 and 1927.[37] By 2012, India had more than 900 million mobile-phone subscribers. Even if the active users at any moment were closer to 600 million, this nevertheless meant that perhaps 1,000 million handsets were in use. And how many lay broken? There was plenty of work for those who could claim to fix cell phones.

MECHANICS OF THE MOBILE

Training institutes mushroomed. Perhaps the best known in north India was the franchised Hi-Tech Institute of Advanced Technologies, established in 2004 in Karol Bagh's Ghaffar Market in Delhi. In 2011 the Hi-Tech Institute had more than two dozen outlets, including one in Nepal. It generated imitators: a host of smaller local institutes that offered similar courses and diplomas, such as Alpha Institute located at the outskirts of Banaras. (See Illus. 17). The typical duration of such a course was about three months at a cost of between Rs 10,000–12,000 (US $200–$240). Teaching was in batches of two-hour classes across the day. It was almost exclusively a male profession. Many of the young students attending Alpha Institute were from the towns and villages of Uttar Pradesh and were the first from their families to move out of agriculture. They viewed mobile repair as an attractive job opportunity for which many were willing to cycle up to 40 kilometres (25 miles) a day to and from their villages. They were responding to the mobile-phone industry's needs and opportunities in similar ways to the young women of the Nokia factory at Sriperumbudur. According to the proprietor of a training institute, all students would find work quickly and with an average salary of Rs 8,000 (US $160) per month. The most successful students advanced to work as engineers in the service centres of brand-name mobiles, such as Motorola and LG, or in the largest and most developed Nokia Care centres.

Raj Kumar found work in one of the three Nokia centres in Banaras. A former student of Alpha Institute, he explained his progress in the industry. After completing training at the age of twenty-one, he gained experience as an employee for a local mobile repair shop. Raj Kumar went on to work for Motorola for nine months, followed by a shorter period at an LG service centre. He subsequently decided to take the entry exams to become a Nokia technician. The exams, he explained, took place over a period of five days where the applicants were presented with the task of repairing 200 handsets. After passing this exam he began work as a junior technician on Rs 10,000 a month. To update the expertise of employees as new models were released, Nokia conducted one-day workshops usually in one of the main hotels of Banaras. Over the next five years Raj advanced to become head engineer, supervising more than thirty junior engineers and twenty 'Care officers' in Banaras' largest Nokia Care centre in the locality of Sigra. Raj said that later he resigned over differences with a superior, and in March 2010 he opened his own shop selling mobile phones and accessories and offering repairs and servicing.

The intermingling of 'informal' and 'formal' sectors extended beyond individuals. All the training institutes that Doron visited offered courses that taught how to break into mobile phones and reformat them. The headquarters of north India's most successful mobile repair institute, Hi Tech, was located in the heart of the grey economy, New Delhi's Ghaffar Market. There was thus a constant flow between the 'formal' and 'informal' sectors. Service centres for brand-name phones relied on such institutions to train potential technicians, who might later work for a multi-national company.

Process

'Mechanics' referred not just to the people who made, groomed and repaired mobile phones, but also to the processes by which hundreds of millions of ordinary people were initiated into using mobile phones and thus to a closer connection with the offices, paperwork and customs of modern states and consumer capitalism. Because of its cheapness and ubiquity, the mobile phone drew people into such networks and tied them to bureaucratic processes that may hitherto have passed them by. Doron's experience with a phone in need of care illustrated some of the choices and processes.

THE CARE CENTRE

In north India in October 2009, Doron's Nokia N2610 phone, purchased in Australia, stopped working. It was unusually hot in Banaras—the temperatures were above 35 degrees—and it was very dusty. On the advice of friends, Doron opened the phone, cleaned the battery by rubbing it fast and hard against his trousers and left it to air under the ceiling fan for a few hours. But the phone did not respond. He was told he had two options: to visit the Nokia service centre, Nokia Care, or seek help from one of the many *mistriis*. While the latter option was tempting, especially in the name of 'fieldwork', Doron decided in favour of the Nokia

Care centre. This was, after all, his personal phone with all his contacts, family pictures, appointments and messages accumulated over years.

Nokia was the leading cell-phone brand from the time the mobile phone arrived on the subcontinent at the end of the twentieth century. Its phones had a reputation for quality, reliability and, unlike many competitors, the ability to withstand the rough conditions of India. Nokia also had a wide distribution network and, like other brand names such as Samsung, LG, and Sony, used Bollywood stars as brand ambassadors in its advertising campaigns. Nokia was everywhere. Not surprisingly, once Doron decided to go to Nokia Care, the rickshaw driver knew exactly where to go.

Banaras has three official Nokia Care centres and the nearest was in a building just off the busy Bhelapur crossing. The two-story building housed the well-known Kerala Café frequented by young people and families looking for a cheap, nourishing south Indian meal. But the shops surrounding the Kerala Café, which used to sell a variety of merchandise from home appliances to shoes and clothes, had been replaced by mobile-phone shops selling all things mobile-related, including memory cards, battery chargers, headphones and even a 'full-body' lamination designed to reduce the wear and tear of dust, extreme weather and the hazards of working people's daily lives. The rapid transformation of commercial priorities echoed those in Saharanpur, 900 kilometres west, that Anuradha Aggarwal described to Jeffrey in Chapter 3.

Arrows on the wall of the staircase directed customers to the Nokia Care centre on the second floor. The walls were adorned with posters, and each floor housed smaller shops with signs in Hindi and English proclaiming that they too were 'authorized service and sales centres' for Nokia. But Nokia Care was the most celebrated of all the businesses. Several large pot plants led to a lofty cardboard archway behind which was a larger-than-life poster of the Bollywood star Priyanka Chopra, sporting her Nokia cell phone and dressed in a glittering evening gown. At the door a security guard doubled as the person who handed numbered tickets to arriving clients to assign them their place in the queue. The reception area was spacious and clean, its white shining tiles giving it a sense of order. All the customers were men, waiting their turn, seated on blue plastic chairs and gazing at the muted flat-screen TV. (See Illus. 18).

The receptionists were the initial contact for the customer. Separated by glass screens, four women registered on their computers the details related to customers and their mobile-phone problems. Behind them was a smoky-white glass barrier, where Doron could glimpse several technicians at work. They too had neatly arranged work stations, at which were laid out instruments to identify problems and repair phones. The customer did not interact with the technicians; all customer care was through the receptionists. Once Doron's number flashed on the electronic board, he approached one of the receptionists, Sita, a young woman in her early twenties who had recently begun working for Nokia in customer care. She listened attentively to his description of the lead-up to his phone's sudden death. As she listened, she typed. She took the phone, quickly examined it and passed it across the counter to a technician who dismantled it for appraisal. This took less than five minutes. Sita gave Doron a form to fill out, which called for Doron's personal details, including name, address and contact number. It specified the cost of the evaluation (Rs 100 or about US $2.00) and an estimated repair cost of Rs 350 (US $7.00).

The roles of women in the service centres contributed to the notions of a 'new India' that glitzy malls and tenants like Nokia Care projected. As receptionists, the women were in positions of some authority; they wore jeans and tee-shirts and addressed customers in formal Hindi rather than the local Bhojpuri of Banaras. This was different from what went on among the *mistriis* in the streets where there were almost no women. Gender, however, also fell into familiar patterns: men did technology; women did meet-and-greet.

Yet the presence of women helped to transform gender assumptions. The pervasiveness of the mobile phone created thousands of wage-earning positions that women filled, and these jobs began to become natural and accepted even in small-town India. The receptionist in a mobile care centre needed to be at least bilingual and fairly well educated. As in the Nokia factory in Sriperumbudur, young women probably cost less and could be seen as less troublesome than men. In a modest way, the need to care for hundreds of millions of cell phones created circumstances a little like those of Rosy the Riveter, the symbolic woman in American industry during the Second World War. The need for workers opened opportunities for women.

MECHANICS OF THE MOBILE

Sita, the young woman who attended to Doron, assured him that the phone would be fixed and ready for pickup by 6 that evening. Doron went next to the cashier, also a woman, for payment; then he lodged a few more forms. On the way out, he noticed a shrine with a photograph of the proprietor's guru, adorned with fresh flowers. Beside it stood a medium-sized, sparkling and well-filtered fish tank with several goldfish. By the door, next to the security guard, a large white board asked: 'Has Nokia made you smile today?' On the upper left corner, a blue-boxed caption proclaimed: 'Nokia: India's most trusted brand'. The board had magnetic yellow 'happy faces' on it, under which customers placed feedback notes, many in English, others in Hindi. Many were satisfied, but others were not so happy with the 'Nokia Care' experience.

Outside, the steamy heat, noise and pollution of Banaras startled Doron back into an older India after what had seemed like a satisfying experience with the new, improved, big-brand India that Nokia and global capitalism had brought. Little did he know that he would be returning to Nokia Care three more times that week. Each time, he was told the phone would be ready tomorrow. On the third visit, Doron took the phone back and presented it to a street-based *mistrii*. In a few hours, the *mistrii* confirmed the Nokia diagnosis that a new motherboard was required—and Doron chose to buy a cheap China Mobile from the grey market in Banaras' Daal Mandi.

Doron's experience dramatised choices forced on an owner of a mobile phone. The sleek Nokia Centre or the dusty turmoil of the bazaar? The sanitised customer-care centres were similar to multinational chain stores all over the world and a testimony to the homogenising effects of globalisation. Banaras, the most famous city of the Hindu faith, got its first McDonald's in 2005. But the spread of global capitalism was not one-dimensional. Homogenisation of practices proceeded unevenly and was shaped by local conditions and preferences.[38] Certain values, ideologies and modes of conduct, however, bore the marks of global modernity. The belief that individuals have equal rights and responsibilities regardless of their place in society was one such idea. In the case of Nokia Care, this translated into the idea of a first-come-first-served, carefully managed, queue.

As mobile phones made wider and deeper inroads into Indian society, engagement with the corporate world spread beyond the upper classes. As well as queues, places such as Nokia Care stressed customer satisfaction and loyalty to a corporate identity. Customers of all classes became socialised into the technical and social practices of a consumer-driven world. These practices required acceptance (if not adoption) of new modes of being in the world, such as embracing the behaviour of the queue, interacting with strangers who had only one transactional role in a person's life and acquiring rights and obligations by contract. This 'individuating' experience contrasted with a more 'relational' mode of being bound by family relations, status, duty and obligations with known individuals, rather than *abstract* rights and contracts.[39] At the most obvious level, the mobile-phone experience was profoundly individual: one person usually talks to another person. But the cell-phone experience introduced another set of mechanisms, which implicated people in a range of transactions: to acquire first a phone, then a service provider, to top up its prepaid talk-time and later, to have a phone fixed when it failed. The experience of Nokia Care conditioned people to 'appropriate' forms of behaviour as consumers.

Middle-class people could use both older methods, found among the *mistriis* and in the grey markets, and the new ways of the Nokia Care experience. Doron recounted his phone-repair experience to a friend named Sunil. The conversation was illuminating.

> SUNIL: When my phone broke I took it to the Sony service centre. The person at the counter told me that to fix this model they'll need to import the part and it will be very costly and would take more than two weeks. I did not have the time or money for this so I decided to take it to the *mistrii*.
>
> AD: What did you think about needing to wait in line at the care centre?
>
> SUNIL: There was a queue there too, but all the places are like this. These companies come from the West, and like any civilized society in the West they have queues, things that need to be organized, with numbers. But the Nokia Care centre is the best one. They are the biggest mobile company in India. Here in India I don't have to queue because relations (*sambandh*) here are more personal. You see, in India more work can be done through personal connections (*Bhaarat mein zyaadaatar kaam niji sambandhon se hota hai*).
>
> AD: What do you mean?
>
> SUNIL: If you have good relations with the shop owner or with the

repair wala, your mobile will be fixed first. If not, then you will have a bit of trouble and you will need to wait until your number comes up.

AD: So when you left the Sony service centre where did you go to fix your mobile?

SUNIL: I went to one boy I know in my *mohalla* (neighbourhood). He deals with second-hand mobiles and also repairs them. He checked my mobile and said he couldn't repair it himself, as it looked like a major problem. So he said he'd take it to Daal Mandi and let me know the problem and cost by that evening. He later called to tell me that it would cost around Rs 700 to repair because it was a problem with the 'motherboard', and it would be ready the next day. My father was upset and said I should not let the *mistriis* in Daal Mandi fix my mobile because they replace the good parts with used ones. But I decided to do it anyway and the mobile was repaired. They replaced the screen and did software formating too. But probably because they had an older version which did not match my mobile, some features like my camera and torch still don't work well.

Sunil is an upper-caste, lower-middle-class man in his early thirties. People of his status have more choice than the poor about where they might take their mobiles for maintenance and repair. They run a risk: once the seal is broken and the handset is tinkered with in an unauthorised setting, service centres refuse to honour a warranty and refuse to fix the phone—as Doron was informed when he mentioned such an option on one of his visits to Nokia Care.[40] Sunil seemed comfortable in both environments. While the branded service centre was his first port of call, he did not shy away from the street-repair economy. People like Sunil were able to benefit from the two seemingly distinct modes of repair: formal and informal.[41]

Service centres socialised consumers and introduced new practices and knowledge in other ways. The receipt that a customer received at the end of a transaction reinforced ideas about contracts. With that in hand, customers could be expected to feel relieved and reassured, guaranteed of Nokia's international standards and confirmed in their own participation in the 'global hierarchy of value'.[42] Even the feedback mechanism, which in the street-side economy was based on word-of-mouth, in Nokia Care was replaced by the notes attached to a board by smiley-faced magnets. Such forms of feedback were intended to give clients a sense of connection in their dealings with an impersonal institution like a multi-national corporation. Having one's mobile repaired in a Nokia Care centre might not have been as quick or convenient as getting it fixed at the local street-repair shop, but it promised certainty and accountability.

The 'choreography of modern consumerism' was attractive to middle-class consumers, though it tended to alienate and intimidate lower socio-economic classes, such as the boatmen Doron knew. They never visited such places. This was not to say they were unaware of what a warranty meant or that they did not own brand-name phones. But the informal sector offered more possibilities, both materially and symbolically, and with fewer restrictions.

India brought its own cultural and material conditions to global fashion and planned obsolescence. Across the world, the average consumer was said to replace a handset every twelve months, and India's middle classes felt similar pressures and temptations of consumer capitalism. But deeply engrained frugality, and widespread poverty, meant that large numbers of fashion-conscious young people turned to India's long-standing practices of repair (and recycling) to satisfy the itch for newer and flashier phones. They embraced the burgeoning second-hand market, especially in the area of smart phones. These second-hand phones could be found in Delhi's grey markets and other more formal shops across the country. The latter even offered limited warranty on second-hand purchases for consumers eager to update to a newer model but unable to buy the latest Nokia, HTC or iPhone. Not surprisingly, this appetite for the second-hand smart phones also made its mark on the internet. Sites for the selling and purchasing of such mobiles, generally handsets no older than a year, became increasingly popular across Indian cyberspace. For many among the poor, the 'unofficial' realm was where the majority of people still went to buy and repair their mobile devices—just as Doron ended up doing.

As Henry Ford discovered in the 1920s, it was difficult to bring the practices of the factory floor to the specific needs of impatient consumers. However, what the automobile industry did in the United States and Europe in the twentieth century was repeated in some ways by the mobile-phone industry in India in the twenty-first. The unpredictable cascade of jobs that a desirable, mass-produced item generated brought new kinds of formal and informal employment to millions of people. The process could

MECHANICS OF THE MOBILE

promote fundamental changes in cultural practices, such as more women in public, dressed differently, interacting with male customers and wielding authority by taking money and validating forms. The mechanics of the mobile also brought millions of people into closer relations with governments, corporations and bureaucracies. It was difficult to have a phone without filling in a form. The mechanics of the mobile included the practices that people had to learn in order to get a phone and make it work. Once consumers had been inducted into such practices, a multitude of uses suggested themselves, as the following chapters explore.

PART THREE

CONSUMING

5

FOR BUSINESS

There is relatively little evidence for the assertion that mobiles help people start new businesses.

Jonathan Donner and Marcela X. Escobari,
Journal of International Development

Celebrations of the mobile phone as a tool for the economic benefit of the poor are common. Sometimes they are well founded.[1] Mass availability of mobile phones, Peter O'Neill predicted in 2003, 'will lead to economic expansion from the bottom up, in part because direct information for marketing can eliminate the urban, middlemen market-makers'.[2] A study conducted throughout India from 2006 to 2008 found that states with higher mobile-phone penetration had more rapid growth in State Domestic Product (SDP). For every increase of 10 per cent in mobile penetration, SDP was said to grow by 1.2 per cent. The study suggested that when one out of every four people in a state had a mobile phone, significantly higher economic growth was achieved. 'The growth dividend of [the] mobile is substantial', the authors concluded.[3] The same study was more muted about the benefits farmers derived from mobile phones: 'mobile phones ... are starting to deliver agricultural productivity improvements'.[4]

The benefits to workers in many old-style jobs are easy to imagine and to see. Vegetable sellers, who for eighty years pedalled bicycles through residential streets crying their wares (and for generations before that had carried the baskets on their own shoulders or on the backs of donkeys), increasingly took orders by phone and delivered to the householder's door. Time, energy and waste were saved. Similarly, day labourers and artisans of all kinds found work—and were found by employers—quicker and easier. (See Illus. 19). But to what extent did the mobile phone really change the capacity of poorer people to work more profitably and effectively? And did mobile phones enable people previously excluded from profitable economic activity to find new ways to earn a living?

A survey of small businesses in a few cities in 2008 found eleven different occupations relying on mobile phones. Two were the products of the phone itself—they used the mobile phone to pass on advertising messages and record information for marketing of products. Another had created a labour exchange for manual workers, based on the mobile phone. A worker with a phone could become a member and thereby be connected with the agency and matched with suitable work. Once connected to the system, workers were helped to sign up for basic bank accounts and to take courses to improve their skills and qualifications. Taxis, rickshaw drivers and waste-paper dealers—random, catch-as-catch-can services in earlier times—had embraced mobiles phones enthusiastically. (See Illus. 23). Examples came from across the country. In a vegetable and fish market in Panaji in Goa in 2010, for example, the owner of a small vegetable stall received orders from regular customers by phone and prepared the order for customers to pick up later. For trusted customers, the owner delivered to the door, and while he was absent, his two employees from Bihar operated the stall. He supplied them with one mobile phone, but he made sure that it could take only incoming calls. There was to be no idle chattering to distant Bihar at his expense.[5]

Mobile phones drew less obvious enterprises into the larger processes of small-scale capitalism, as the case of Ranjeet Gupta, a henna artist, vividly illustrated. In north India, Pakistan and north Africa, women inscribe elaborate patterns on their hands and feet as decoration before auspicious events such as weddings. Henna is the plant that provides the coloured powder and paste used to draw the patterns. Gupta began his career on the pavements of west Delhi in 1999 where poorer women would sit on a stool while he decorated their palms for a small sum. 'He was', the researchers noted, 'harassed by officials since pavement businesses were technically illegal, although popular'. In 2002, he got a cell phone (when the researchers met him in 2008, he had two). 'His clients now call him'. His business by this time involved his being booked to visit people's homes to apply henna; the customers were wealthier and the designs more elaborate. When the researchers met him, he had also gone into the training business and offered five-week courses to aspiring henna artists for Rs 3,000 (a month's salary for a basic low-end wage earner at the time).[6]

The tiny businesses that embraced the mobile phone drew their owners deeper into the arms of the modern state. To begin with, one could not legally get a SIM card without filling out a form, attaching a photograph and being provided with a phone number.[7] (See Illus. 20). The form itself required the applicant to provide a permanent address as well as personal details that the service provider was supposed to verify. These requirements also applied to the handset which had to have an IMEI registration number. After the attack on Mumbai by phone-carrying murderers in November 2008 (Chapter 8), these requirements were applied more strictly, and users and companies who tried to circumvent them were subject to heavy fines.

Many users of mobile phones had never had a bank account. Getting a phone, however, gave them an identity number through their phone, linked them to bureaucratic systems of record-keeping and forced them to keep track of their use and their phone payments. The phone was disciplining or educating them about the requirements of modern worlds. Their movements could also be monitored, and they in turn could keep track of others more closely than ever before. Homesick Bihari employees, for example, could yearn to make a call back home, and, at the same time, they could be checked up on by their absent boss whenever he wished.

None of these processes is new. For 200 years, as rural Indians have been absorbed into the offices and factories of expanding cities, people have been exposed to bureaucracy, discipline and homesickness, along with widened opportunities and new perils. The cheap pocket watch, and later the wrist watch, were handy devices for enforcing punctuality on workers. The cheap mobile phone, however, made things happen faster, more widely and more intrusively—and for more people—than any similar personal device. And it let people talk back and reach out as never before. The mobile phone quickly became an integral part of everyday life, especially for those small businesses that had little access to wider markets and whose business often entailed danger, such as the fishermen of Kerala.

On the sea ...

For a time, they may have been the most famous fishermen and women in the world, the low-caste, hard-scrabble fisherfolk of Kerala who began to appear in journalists' accounts and economists' research papers from about the year 2000. Fishing in Kerala was a low-status occupation: fisherfolk were 'historically branded as outcasts'.[8] The people who fish, whether Hindus, Muslims or Christians, were seen as working in an unsavoury occupation. They killed living creatures, dealt in a smelly product and faced great risks from bad weather and the perils of the sea. They were early adopters of the mobile phone, and their story illustrated the technology's possibilities to make lives safer, surer and more prosperous.

One of the first of these stories appeared in the *New York Times* in August 2001. 'In the seas ... off the coast of southern India', Saritha Rai wrote, 'the steady drone of motorized fishing boats is often interrupted by the ringing of mobile phones'. Rai described how the owner of a trawler based near Kochi in Kerala had got a mobile phone a year earlier and 'doubled' his profits by being able to compare prices at different harbours before landing his catch. 'Life without a mobile phone', her informant tells her, 'is unthinkable'.[9] Another American journalist, Kevin Sullivan, described his time on a 74-foot trawler off the Kerala coast a few years later. The captain dropped his nets, and within minutes his cell phone began to ring. Agents onshore had heard that he had found a shoal of fish and were ringing to bid for the catch. 'When I have a big catch', the skipper told Sullivan, 'the phone rings 60 or 70 times before I get to port'. The journalist concluded that the cell phone 'meant greater access to markets, more information about prices and new customers for tens of millions of Indian farmers and fishermen'.[10]

FOR BUSINESS

From 2001 Kerala's fisherfolk and their mobile phones figured in dozens of studies and newspaper reports, including a seminal essay by Robert Jensen, an American economist. Published in 2007, Jensen's study lent academic rigour to the widely discussed impressions that mobile phones were particularly good for fishermen. Jensen found that from 1997 when mobile phones were introduced in Kerala, the construction of transmission towers very soon allowed mobile signals to be received up to 25 kilometres (15 miles) out to sea, and within four years—by the time Rai wrote her piece for the *New York Times*—'over 60 per cent of fishing boats and most wholesale and retail traders were using mobile phones to coordinate sales'.[11] The results were better prices for fishermen, more stable prices for consumers (which in the past had fluctuated widely when some landing areas received more catches than there were buyers and others, more buyers than fish), and 'the complete elimination of waste'.[12] 'Overall', Jensen concluded, 'the fisheries sector was transformed from a collection of essentially autarkic fishing markets to a state of nearly perfect spatial arbitrage': that is, the mobile phone allowed fish to find buyers and buyers to find fish. Profits increased yet prices fell.[13] By 2008, it was estimated that up to 100,000 people associated with fishing in Kerala had mobile phones.[14]

Jensen's study was picked up by researchers, journalists and teachers around the world, including C. K. Prahalad's best-selling *The Fortune at the Bottom of the Pyramid* and the hard-headed British weekly the *Economist*.[15] The latter welcomed the study for adding rigorous research and grassroots experience to a wealth of anecdotes and a number of macro-studies about the economic benefits of mobile telephony. The best known of the big studies suggested that an increase of 10 per cent in the penetration of mobile phones in a region added about 0.5 per cent to Gross Domestic Product.[16] Working people quickly realised the benefits, the *Economist* contended, and 'fishermen, carpenters and porters are willing to pay for the service because it increases their profits'. And it all happened without governments. Mobile phone networks were run by private operators 'because they make a profit'.[17]

Kerala's fishermen offered a particularly picturesque example of the positive aspects of the mobile phone. (See Illus. 21). A beautiful coastline, a nutritious product and a dangerous industry provided the background for accounts of the benefits of the technology. From the perspective of a fishing boat on a choppy sea, a fisherman told a reporter that 'the two crucial changes' had happened to fishing in his lifetime: 'the inboard motor and the mobile phone'. From the ivory tower of business studies, C. K. Prahalad said that 'one element of poverty is the lack of information. The cell phone gives poor people as much information as the middleman'.[18] As well as allowing fishermen at sea to discover where best to land their catches, their mobile phones kept them informed of changing weather conditions, allowed them to alert friends on other boats to substantial shoals of fish and enabled them to call for help if a boat ran into trouble or to head for shore if there were family emergencies on land.[19]

The dirt, the dangers, and the disparities of fishing in Kerala, and elsewhere in the world, have been well known for a long time. In *Chemmeen* [prawn], the most famous novel and film about Kerala's fishing people, the role of middlemen as exploiters of poor fisherfolk is central. '"You can take your fish back!" the big trader said', when the desperate fisherman tried to sell at an unfavourable moment.[20] Fishing in Kerala provides a parable for many small-scale occupations that embrace the mobile phone. The mobile improves the odds for poor people against power-holders; but it does not even the odds. Take, for example, the boat that the *Washington Post* journalist Kevin Sullivan went to sea on: a 74-footer (22.5 metres) with a crew of more than 30.[21] The boat represented a huge investment by the standards of ordinary fishing families along the coast, and even the cost of diesel to run its engines amounted to many days' labour for old-style fishermen working from small catamarans. Mobile phones allowed big boats to cover more territory at sea and scoop up more fish. The increased safety provided through weather reports emboldened big boats to move farther afield even during the monsoon. Fishing during the monsoon had been a controversial issue on the Kerala coast from the 1980s when movements of small fisherfolk, sometimes led by activist Catholic priests and nuns, forced governments to limit perceived overfishing. By these measurements, however, the gap between big boat owners and smaller players could be seen as remaining the same, if not widening, with the arrival of mobile phones. And the increased intensity of fishing could deplete the long-term stock of fish.

Caution and nuance are essential if we are to understand the downs as well as the ups that the technology brings. The tales of the Kerala fishermen, vivid for their positives, also provide apt cautions against going too far in celebration of the potential of mobile telephony. The autonomy of the mobile made fishermen of all sizes safer, less wasteful of their catch and less exposed to wide fluctuations in price. The mobile also, according to one account, reinforced long-standing tendencies among the men who go to sea to cooperate rather than compete. 'The availability of mobile technologies', Sreekumar writes, 'has amplified this impulse and enabled new modes of cooperation'.[22] The key point of course is that the technology reinforced and extended practices, rather than created them.[23] The mobile did not alter class relations, yet much of the writing about cell phones and fisherfolk overlooked the difference between the 74-foot diesel-powered trawler fishing 20 miles offshore and the two-man catamaran working within sight of land. The phone allowed the owners of big boats to fish more widely, efficiently and constantly, and the toll of increased fishing on fish stocks was hotly debated. Similarly, though the phone allowed price comparison at different harbours while still at sea, it also allowed traders on shore to keep in touch with each other, maintain solidarity and keep prices low.[24] The power of the middleman 'has lessened somewhat... the fishermen get the opportunity to drive a harder bargain than before'.[25] The mobile phone improved some conditions; it did not reorder society.

Around the globe ...

The fisherfolk of Kerala provided dramatic, widely known examples. Evidence from around the world confirms aspects of the Kerala story but adds the rider that culture and practice adapt technical innovation to local ways. The American scholar, Jonathan Donner, one of the outstanding analysts of mobile technology, and his colleague Marcela Escobari, provided a revealing examination of mobile telephony and small-scale businesses around the developing world, ranging from the Caribbean and Africa to India. Two of their findings are particularly significant for attempts to understand the impact of mobile phones for small enterprises in India.

First, the Donner and Escobari inquiry found that mobile phones seldom led to the development of *new* businesses. Much of the impressionistic writing about mobile phones tended to see the phone as a key by which people previously excluded from ways of earning their livings were enabled to do so. The evidence, however, generated a string of negatives. Research did not confirm uplifting stories of new jobs created by new people. Just as there was no evidence that people who were not previously fisherfolk got into the fishing business in Kerala when mobile phones made it safer and more efficient, new people had not created *new* enterprises elsewhere. 'There is relatively little evidence for the assertion that mobiles help people start new businesses'.[26] Another of the veteran authorities on mobile telephony was similarly surprised in a study in Jamaica. 'The phone is used much less amongst low-income Jamaicans in connection with either jobs or entrepreneurship than we anticipated', wrote Heather Horst and Daniel Miller.[27]

A second aspect emerged from the research of Donner and Escobari. The early accounts of Kerala's fisherfolk suggested that the mobile phone allowed producers—the people who caught the fish—to undermine the control that traders and middlemen exercised over prices. In the past, traders benefited from the fact that fish were perishable and a catch, once landed, could not easily (or fragrantly) be taken to another market. However, when Robert Jensen did his research soon after the mobile came to Kerala, the phone gave fishermen the opportunity to negotiate with buyers on land while they were still on the water and thereby determine where they would land their catch. Within ten years, however, 'the collusive behaviour of the buyers' had 'become a gross material reality'.[28] This was also the experience elsewhere. 'None' of the studies reviewed by Donner and Escobari suggested that 'mobiles help MSEs [Micro and Small Enterprises] bypass middlemen'.[29] Nor should this be surprising. Traders began with the advantage of better access to communications and information than producers. (Recall Chapter 1 and the fact that in 1761 the merchants of Pune knew of the Maratha defeat at Panipat days before the political rulers and the general populace.)[30]

These pictures were less exhilarating than some more impressionistic and optimistic writing about telephony had conjured up. But the 'no new work and no improvement in equity' picture needs to be modified in one significant way. New jobs *have* resulted in the area where one would expect: in the support,

FOR BUSINESS

expansion and imaginative development of mobile telephony itself. The growth of a mass-based industry, as we saw in Chapters 3 and 4, needed hundreds of thousands of people to service it, just as the railway and automobile industries created huge demand for workers in the nineteenth and twentieth centuries.[31]

Two examples illustrate how making a living in India changed with the advent of mobile technology. The first is a story (with an uncertain ending) of how technology enabled economic activity that was previously impossible: simple, mobile-phone-based banking, similar to M-PESA in Kenya, but with distinctive Indian features. The second example analyses the general phenomenon, noted by Donner and Escobari, of the mobile phone improving productivity and generating 'changes *in degree*' rather than 'changes *in structure*'.[32] The boatmen of Banaras, like the fisherfolk of Kerala, engage in an occupation almost as old as the River Ganga itself; but the mobile phone—and the diesel engine—changed some of the ways in which they plied their trade and conducted their lives.

At the bank ...

In 2009, as M-PESA was passing the six-million-client mark in Kenya, the brothers Abhishek and Abhinav Sinha launched EKO in New Delhi. The idea was similar to that of M-PESA but geared to Indian demography and practices. The brothers were from Bihar, both had graduated from the Birla Institute of Technology near Ranchi in Jharkhand state and Abhishek had been a founder of a successful company that created software to improve the process for topping up mobile phone accounts electronically. 'We were doing an application for recharging talk time for an operator in Oman', Abhishek Sinha explained. Strolling and talking, he and his brother passed a 'small *paan* kiosk. Just imagine', they said to each other, 'that the guy who wants to buy a cigarette can do it with his cell phone'.[33] Their thoughts were in tune with other engineers and economists who sought ways to digitise money—to create 'mobile money' or put a cash card on one's phone.[34]

M-PESA (M for Mobile and Pesa for 'money' in Swahili) differed from EKO significantly. M-PESA was run by a mobile phone operator, Safaricom, which was in turn a subsidiary of Vodafone. EKO was an independent enterprise, owned neither by cell-phone

123

providers, nor by banks. But EKO developed formal arrangements with the State Bank of India (SBI) and the ICICI Bank as a 'business correspondent', which gave its depositors guarantees and services and allowed EKO to describe itself as providing banking services. It also allowed it to place the reassuring logos of the State Bank of India and the ICICI Bank on its posters and advertising materials.

India presented tantalising challenges. Policy-makers agreed that the state, for all its proclaimed efforts to banish poverty (*garibi hatao!* had been a rallying cry from 1970), had too often failed to get resources into the hands of the people for whom they were intended. This resulted from India's citizens not being enmeshed effectively with the state—through tax-paying, school attendance, electricity and water connections, medical provision or bank accounts. It was one thing to say in effect, 'We're the state and we're here to help you', but it was another to connect with the people that the state intended to help. The cheap, mass mobile phone appeared to offer a way. But could procedures and technologies that worked for a few thousand people—or even a few hundred thousand—be made to work for hundreds of millions? The problem for the Sinha brothers and others like them was to create a method of banking that was no more difficult than topping up one's pre-paid mobile phone account.

In developing a banking model for India based on mobile phones, a key requirement was to instil trust. Why would someone hand over their hard-won savings without having a printed receipt or a passbook? An answer appeared to lie in the mobile-phone experience, and in bringing into the EKO system local shopkeepers who were already part of the networks selling Fast Moving Consumer Goods (FMCG). Paperwork and bookkeeping deterred many semi-literate users of cheap mobile phones; well-known local shopkeepers had been essential in overcoming those fears. Familiarity, simplicity and trust were three essentials in inducing poor but busy people into new practices, particularly those that risked meagre savings. 'The choice of us using a grocer as a community banker', said Abhishek Sinha 'is that he is already a trusted guy'.[35] To get a mobile phone, people needed to do only minimal paperwork at the sign-up stage, and EKO aimed to make getting the basic facilities of a bank just as simple. But to achieve trust and simplicity, efficiency was vital. Anupam Varghese,

EKO's vice-president for research and design, described this as a process of 'learning and unlearning'.[36]

What were the virtues of having a bank account? Once people had one, they were in relations with the state, which could monitor them in various ways, yet a bank account also allowed them to keep their earnings safe, to save money and to pay bills or aid relatives. The disadvantages of people who did not have bank accounts were well illustrated in the difficulties in transferring wages to workers under the Mahatma Gandhi National Rural Employment Guarantee Act (MGNREGA), which guaranteed 100 days' work a year at a minimum wage to any Indian citizen who asked for it. Begun in 2005, MGNREGA was the largest scheme of its kind ever attempted, and one of its greatest challenges was to ensure that a worker's wages ended up in the worker's hands without intermediaries siphoning it off. The program also aimed to give every worker a bank account to improve the chances of full wages finding their way to rightful recipients.

Surveys found that bank accounts did not eliminate corruption, but they helped. Even illiterate people, who were 45 per cent of the sample in one survey, had 'an interest in learning how to use the banking system', and more than 75 per cent of the sampled workers liked receiving their wages via a bank. But distance was a problem: more than 40 per cent had to travel more than five kilometres to a bank, and a number had to travel more than 10 kilometres, effectively losing a day's pay in order to withdraw wages.[37] Getting a bank account was intimidating. In the sample, 60 per cent of workers went with either an official or a contractor to open an account. Paperwork had to be learned, and corrupt and crafty intermediaries could induce workers to sign piles of withdrawal slips or manipulate their passbooks. There was nothing new about the intimidating quality of banks. 'When I go into a bank I get rattled', the Canadian humourist Stephen Leacock wrote of small-town Ontario in the 1890s:

> The clerks rattle me; the wickets rattle me; the sight of the money rattles me; everything rattles me. The moment I cross the threshold of a bank and attempt to transact business there, I become an irresponsible idiot.[38]

Leacock had a PhD in Economics from the University of Chicago. Illiterate people in twenty-first century India were more enterprising than Leacock (who vowed in future he would keep his savings in silver dollars in a sock). 'For now I am learning', one MGNREGA wage earner explained about going to the bank with a helper. 'Later I will go on my own'.[39]

Indians in the first decade of the twenty-first century were under-banked and widely dispersed. Seventy per cent of the population lived in 600,000 villages, but there were fewer than 75,000 bank branches, heavily concentrated in cities and large towns. By 2012, however, there were more than 900 million mobile phone subscribers.[40] If mobile phones could become mobile passbooks, greater honesty, efficiency and equality might be achieved. People were willing to learn, and mobile phones were becoming cheap enough for even the very poor.

The challenge for an organisation like EKO, according to its founders, was to develop a system that would make sense to poor people. But who were 'the poor', what did they need and what could 'mobile banking'—whatever that might be—do for them? The Sinha brothers and their colleagues went back to Bihar, one of India's least prosperous states, to learn about people's everyday practices and needs. For many Biharis saving money was practically impossible because of the crowded conditions of the joint-families in which they lived. Varghese said:

> People did not have a safe place to store and save money. [This was] a concern voiced by women whose primary chore was to manage the household economy, yet they were often harassed to give what meagre savings they saved to their husbands who were often unemployed and took to the drink.[41]

There were also technical aspects that required EKO to unlearn some of its assumptions. The EKO team observed how mobile phones were obtained, maintained and recharged. Most people used cheap basic phones; multimedia and smart phones were costly and less common. The neighbourhood shopkeeper—who sold a host of products as well as the recharge coupons for mobile phones—proved to be pivotal for mobile-phone transactions. As trusted, worldly-wise players in the cell-phone world, they helped marginally literate people manage their mobile phones. (See Illus. 22). Shopkeepers knew which buttons to press at what point—should it be a '#' or a '*'?—and had an interest in helping: they got a commission on every card or mobile-phone service sold.

Poor people, though illiterate or semi-literate, had already shown the skills, including everyday numeracy, to operate mobile phones, and using a mobile phone often enhanced those skills. People became familiar with the phone's keypad, learned to follow numerical patterns and became adept at recharging phone credit and checking their balance of remaining talk-time. Such observations led EKO to develop a simple Mobile Money Transfer (MMT) method, as Varghese explained:

> We had to follow a few basic principles to accommodate the lower denominator:
> 1. that our operation could work on low-cost devices
> 2. [that it was able to] make use of the dialling pad interface, as was commonly used for checking one's balance [and]
> 3. [that it was able to] make use of the existing infrastructure, which people were now using for recharging their phones.
>
> Another important practice in India is missed calls. If the Philippines is the SMS capital of the world, then India is the missed-calls capital, and we soon realized that tapping into a missed-calls practice would be far more sensible than trying to do the transactions through a messaging system. Even SMS which seems to us an easy operation meant an additional layer of complexity, and demanded the actual navigation of the keys. Again, we had to 'unlearn' and ensure our method could work across various mobile models on the simplest key pad.

'Unlearning' meant avoiding new practices, even though it seemed essential to have some sort of 'smart' technology for mobile-phone banking. Indeed, according to Varghese, a few companies did attempt to use extra devices for their e-banking schemes, such as biometrics, or smart cards and thumb-impression machines, or even CCTVs, but these did not work well. EKO's view, as its founders emphasised, was to use what people already felt confident with—both the physical infrastructure and social practices—from small shops to routines of mobile phone recharging. If carrying one's mobile phone had become as easy as slipping on one's sandals, using it to gain the services of a bank needed to be as familiar and unchallenging as acquiring new footwear. Operating an EKO-style account was like writing a cheque or withdrawing money from an ATM. The shopkeeper functioned as the ATM; the mobile phone was the debit card; and deposit of funds followed the same logic as purchasing a pre-paid coupon for a mobile phone.

EKO was by no means the only initiative attempting to use mobile phones to provide simple banking to relatively poor people. But the EKO model captured the potential of the cheap mobile phone and illustrated why having a bank account was important for people who had to deal with a modern state and were entitled to benefits from it. By the end of 2011, EKO had 170,000 account holders in New Delhi, Bihar and Jharkhand and more than 1,300 Customer Sales Points (CSP).[42]

These neighbourhood shopkeepers already sold pre-paid services for mobile phones. To register an EKO account, a person followed procedures similar to registering a SIM card. Proof of identity—a ration card or a voter-enrolment card—and a photograph had to be presented. While the transactions themselves were paperless, EKO distributed a small kit to account holders, containing illustrated explanations and, most importantly, a booklet which listed a series of pin numbers. These provided account holders with security over their transactions.

How did it work? After registering an account with EKO, if a client wished to deposit Rs 100, she gave Rs 100 to the shopkeeper and waited while the shopkeeper did the following on the shopkeeper's own phone:

Step 1   keys in <*543*>
Step 2   followed by the <depositor's mobile number*>
Step 3   followed by the <amount to be deposited*>
Step 4   followed by a <10-digit, one-time-only PIN number from the shopkeeper's set of unique numbers registered against his or her name and including the shopkeeper's own unique personal ID#>
Step 5   presses <CALL>
Step 6   the shopkeeper and the depositor at once receive an SMS recording the transaction

To withdraw money, or send money to a recipient with an EKO account, the procedure was the same, except that mobile phone number of the person receiving the money became part of the transaction.

EKO identified existing needs and aimed to satisfy them more effectively.[43] As with M-PESA in Kenya, the majority of customers used EKO services for remittances. In Delhi, for example, migrant workers from Bihar led the way.[44] The mobile-phone system allowed migrants to bypass costly, cumbersome processes that previously hampered their ability to send money home. The system avoided the daunting task of opening a bank account and reduced the uncertainties and charges involved in transferring money through the formal banking system: paperwork, delays due to bank clearances and the time taken to go to a bank branch and wait to be served. Unlike bank branches, local shops were often open for long hours and on holidays. Transfers were immediate: once money was deposited the person received a text message and could draw cash.

Maintaining the system was more complex. A shopkeeper needed to have cash on hand to pay to EKO customers who made withdrawals. Shopkeepers were supplied by couriers who replenished cash whenever necessary, but couriers had to be trained, trusted and supplied with cash from old-style bank branches. Shopkeepers' own balances of EKO-related funds, and their commissions (0.3 per cent of every transaction) from EKO, were maintained centrally and accessible to shopkeepers through their mobile phone. Because all transactions were electronically registered, errors and the potential to siphon off money were minimised. Shopkeepers received a half day's training in the procedures; support and advice were available at the EKO end of their mobile phone.

Basic banking through the mobile phone brought the advantages of a bank account to large numbers of people, previously too poor, unlettered or frightened to confront the forbidding formality of a bank office. For migrant workers, these advantages included, as they had for migrants who flocked to M-PESA in Kenya, the ability to send money home quickly, cheaply and safely and to keep earnings safe, not tied in a handkerchief or hidden in a box under a bed. A bank account also gave poor people, who were entitled to payments from governments or other sources, a secure and prompt way of receiving their dues. This promised to reduce the ability of paymasters who handled cash to demand bribes in exchange for handing over entitlements. At the same time, such arrangements could disrupt long-standing family patterns if payments to a woman, which once came in cash and were seized by her husband or his relatives, now came to an invisible mobile-phone account in her name.

Mobile-phone banking could be disturbing, and could provoke resistance, as a project in the state of Bihar illustrated. When EKO and the Norway-India Partnership Initiative (NIPI), introduced Mobile Money Transfer (MMT) in January 2011 for women who worked as Accredited Social Health Activists (ASHAs), household tensions arose. Some husbands banned their wives from having mobile phones (Chapter 7). Yet the benefits of such a system of payment were clear. For the women who worked as health-care providers, receiving and saving their salaries had long been a problem, though the importance of timely payment was crucial for their household budgeting. In poor households, delays in payment had economic, social and evening-meal implications. They compromised a woman's status:

> Taking up a position of ASHA was not easy for [the women]. Many wanted a relief from the crushing poverty due [to] unemployed/alcoholic husbands... Some took on the job to augment the family resources... many had to face a lot of opposition and had to struggle to convince the family. Essentially the objections were based on the fear that she may become too independent, may interact with men outside the family; [and the] work load will fall on the mother-in-law if daughter-in-law goes out to work, among other things.[45]

For women whose families allowed them to go out of the house and work as ASHAs, regular payment showed that the risk to a woman's reputation if she moved about independently was worth taking. Families, however, needed to be convinced:

> The main issue was suspicion that ASHAs may contact other men, misuse the freedom, will get unwanted calls. Also the perception was that when a man has a mobile he will use it for making money, where a woman may just chat and use up the call time.[46]

By transferring payments through the mobile, women accessed many of the facilities of a bank account, including the ability to check their balance, transfer money and receive interest on their account at four per cent a year—all without leaving home.[47]

The simplicity of the phone was one of its advantages. Close to 70 per cent of the women in the survey got their first phone when they became an ASHA. They told the researcher that the

> mobile phone ... gives us a bit of autonomy and has enabled us to take independent decisions away from the collective family decisions to some extent. The ownership of phone has made it possible for us women, with little or no land ownership, to have a sense of freedom and identity [sic] albeit limited.[48]

More than money was involved. Relations within families were disrupted in ways that many people would see as positive. As regular payment began to arrive without long waiting times in offices or bribes paid to cashiers,[49] opposition to women's work and women's phones subsided in some families. Chapter 7 explores the new, almost-daily transactions and decisions that phones forced on families.

EKO was not alone in searching for ways to use the mobile phone as a poor (or rich) person's bank account. By the end of 2011, phone manufacturers, service providers and banks were exploring ways to use the mobile to get banking to the poor, and make a profit using the Vodafone M-PESA mobile banking model. Nokia, the handset maker, started services in Pune, Chandigarh and Nasik, all prosperous large cities.[50] The two great service competitors, Bharti Airtel and Vodafone, launched phone banking in partnership with the State Bank of India (Airtel) and the ICICI Bank (Vodafone).[51]

The challenge lay not so much in technology—putting basic banking services onto mobile phones—but in the logistics of finding thousands of outlets to act as mini-banks, training their owners and keeping them supplied with cash and information. The challenge was labour-intensive. 'Retailers are very street-smart people and it takes us half a day' to train them as EKO agents, Abhishek Sinha said. But even in its early stages in 2010, EKO had to have twenty of its own fulltime employees liaising with FMCG agents who in turn had hundreds of employees who visited and supplied small shopkeepers.[52] Vodafone by 2011 had more than a million outlets dispensing Vodafone products of various cost and complexity, and the company aimed to turn these agents into banking outlets. Airtel similarly had 500,000 outlets that it aimed to train and supply.[53] 'Upscaling'—increasing the number of customers—depended on increasing the number of outlets whose owners had to be trained, supplied and informed. More than half of the Indian population did not have bank accounts. To put a mobile-phone banking agent within reach required hundreds of thousands of outlets and tens of thousands of people. The people linked the outlets to the formal banking system by delivering cash, PIN code documentation and marketing information.

On the river …

Located on the sacred River Ganga, the city of Banaras (Varanasi) has been a pilgrimage centre for thousands of years. Unlike the ocean-going fishermen of Kerala, the majority of river boatmen in Varanasi relied on passengers, not fish, for their livelihood. Pilgrims and visitors were the main source of income, and the boatmen's boats were their means of production. Mobile phones caused ripples on the Ganga, improving the lives of some and complicating the lives of most, as Sujit's story illustrates.

Tourism and pilgrimage were big business in Banaras, with groups of pilgrims constantly flocking to the riverfront to worship and bathe. Whether they came to the city by bus, train or plane, visitors reached the riverfront mostly by roads which led to the major ghats—the steps down to the river's edge—such as Dashashvamedh Ghat, Assi Ghat or Raj Ghat. Different ghats had different earning potential for historical, cultural and architectural reasons, and Dashashvamedh Ghat maintained supremacy over all other ghats. Its central location and religious significance made it a pivotal point in the city. It would be misleading to speak of Dashashvamedh Ghat as one ghat, as its territory consisted of several distinct ghats. The main road leading to the ghat area branched out into several lanes, each leading to a different set of steps descending to the ghats. Most ghats were only accessible by foot or scooter via the small alleys and were relatively insignificant; pilgrims and tourists were few. On those ghats, resident boatmen earned less and were poorer than their peers on the large popular ghats.

Sujit was a boatman from the small Nishad Raj Ghat who often worked as 'driver' (*mallahi*) for a resident boatman (*ghatwar*) from Assi Ghat. When Doron met him in 1999, this was a common practice. On most of the major ghats, the 'drivers' rowed and maintained the boats of the resident boatmen (*ghatwars*) and received 50 per cent of the earnings. There was no binding contract between the *mallahis* and the *ghatwars*. Even then, some of the *mallahis* shifted from one ghat to another according to demand, and they were often employed during the busy times of the year on the major ghats where there was a large flow of passengers.

With the arrival of the mobile phone, some boatmen were able to neutralise the advantages of the *ghatwars* who owned larger boats on more popular ghats. Sujit became one such beneficiary. Longstanding conventions had decreed that when a visitor entered the territory of a particular ghat, boatmen immediately staked their claim to the person.[54] By making a call (*boli*) the *ghat-*

war had the right to approach the potential passengers to offer a boat ride and negotiate a price. Only *ghatwars* were allowed to participate in this bidding system, and a *ghatwar* had to be present on the ghat to do so. A moral economy underpinned this system: subsistence was guaranteed to all boatmen on the ghats through norms of reciprocity and redistribution. Poorer boatmen could rely on the dominant and more prosperous classes to ensure a minimum income during desperate times. Some boatmen, however, had always been frustrated with the system, arguing that it favoured owners and workers on the major ghats and limited social mobility and initiative.

A key exception in the old system lay in the category of 'known passenger' (*parichit savaari*). These were people with whom a boatman could claim a previous relationship and whom he had the right to pick up wherever they appeared. The conventions relating to the daily tourist and pilgrim traffic did not apply; a boatman could collect his established client unchallenged. The mobile phone allowed a boatman to forge relationships with an ever-increasing pool of potential passengers and undermine the old customs, as Sujit explained.

In May 2009 Sujit looked after a group of pilgrims from the western state of Gujarat and established particularly good relations with them. While they were in Banaras, he gave them his mobile phone number and told them that if they needed anything or got lost in the alleys that lead to the major temples, they could call him. More than a year later, Sujit received a call from one of the group informing him that another party from the same region of Gujarat was planning a pilgrimage to Banaras. He requested Sujit to make arrangements for their boat rides. Sujit said he would meet them at the railway station. A small boatman from a minor ghat was becoming a travel agent.

However, as a boatman from a lesser ghat, Sujit owned a small boat which could barely carry six passengers. To cater for fifty Gujarati pilgrims he had to ask a relative from Raj Ghat, one of the major ghats on the north of the riverfront, to borrow his boat in return for 50 per cent of the earnings. Overall, Sujit made a considerable sum of money from the Gujarati tour party, not only from the boat rides themselves, but from ancillary activities associated with pilgrimage, such as taking pilgrims to sari shops from which he received commission.

Over time, Sujit began accumulating a 'known passenger' list of clients on his mobile phone. He developed a record system—the phone's address book. Pens and paper were not part of Sujit's work tools, and boatmen of earlier eras were often illiterate. But Sujit's mobile phone became a data-base (though he did not call it that) on which he kept the phone numbers of rickshaw drivers, tour guides, touts, priests and shops to which he took customers. He thus began to undermine what he viewed as the unfair distribution of resources that benefited the larger ghats. By establishing direct contact with pilgrims and tourists he leapt some of the barriers of the unpredictable tourist and pilgrim trade, such as the inaccessibility of his own ghat and the system that brought the popular ghats the lion's share of the 'random' work. Sujit's earnings increased.

Mediators still mattered, but boatmen like Sujit now had better access to them through their mobile phones. Though such relationships existed in the past, the mobile phone allowed a boatman to expand his networks and keep them organised. The phone sometimes enabled him to coordinate visits, get more work and earn more money in a day. Boatmen with fewer resources and from smaller *ghats* were able to spread their web further, arrange passenger flow more predictably and no longer be limited to the 'leftover' passengers trickling down from the main ghats. The boatmen exemplified aspects of the 'network societies' written about by scholars like Manuel Castells, David Singh Grewal and Barry Wellman and alluded to elsewhere in this book.

Such changes needed to be seen in perspective. Sujit was a privileged resident-boatman, not a mere 'driver'. As such he had rights to own and ply boats from his designated ghat, even if his small ghat had a very limited flow of passengers. Nevertheless, Sujit had the potential to expand his trade with the help of the mobile phone. The 'drivers' too could benefit since they were able to get more work if they could be reached easily when jobs came up. Yet the nature of the boating industry was such that most drivers were connected to a specific resident-boatman and territory, as the work demanded a degree of loyalty and constant presence on the ghat in case a stray passenger arrived. The socio-economic structures of the riverfront economy constrained the movements and earning capacity of the drivers. Moreover, their bosses—resident-boatmen—were able to maintain a close watch on them by using the mobile phone to check on their actions and whereabouts. The mobile phone could increase one's autonomy, but at the same time impose new limitations. To appreciate the ambiguous effects of the mobile phone, we need to take account of structures that frame the possibilities for action, and which facilitate or constrain the autonomy of individuals, as the following example illustrates.

For Sujit, the changes brought about by his mobile phone were patchy and piecemeal, not sudden and overwhelming. It would take a long time for him to amass the capital to buy a large enough boat for big groups of pilgrims. He had, however, diversified his clientele and created relationships with both foreign and Indian tourists, many of whom passed on his number to friends on the tourist trail. They in turn contacted Sujit on arrival in Banaras to seek his advice and services. These flow-on effects of mobile communication were hard to measure precisely because they constituted only one variable in a network of socio-economic considerations. Through the mobile phone, Sujit and other boatmen expanded their networks, created personal ties and coordinated relationships with potential customers and partners. They began to navigate a system that in the past heavily favoured powerful boatmen on larger ghats. The mobile phone by no means eradicated the differences, but it put Sujit on a slightly more equal footing.

Such 'equalising' brought costs and generated tensions. As Sujit's story illustrated, when technology enters a social structure, people make that technology work—they give it meaning—within the existing structure. The mobile phone extended the category of 'known passengers', which the boatmen had long acknowledged. A boatman could develop a list of 'known customers' on his cell phone and arrange to meet them without having to organise meetings face to face. An enterprising boatman from a less frequented ghat could thereby expand his business. But tensions surfaced as competing boatmen sought to safeguard their interests against a technology that potentially undermined their earning capacity.

Boatmen from the larger Assi Ghat spoke of this as a damaging consequence of mobile phones. They expressed frustration at the fact that their ghat had turned into a pick-up point for empty boats coming for passengers who had made arrangements by phone with boatmen from other ghats. Their view was that this practice by 'greedy boatmen' from other ghats breached the age-old system of regulating passengers. How these conflicts would be solved remained an open question, but it highlighted one of the disruptive (and potentially explosive) effects that mobile phones had on community life and business practices.

By 2009, boatmen said that mobile phones—something unknown when Doron first met them ten years before—were essential for their business. All of them used the phone as an alarm clock to wake up at dawn, the most popular time for boat rides among tourists wishing to experience sunrise on the Ganges. Many boatmen noted that they rarely turned off their device, because clients could call at any hour to book their boat. Ramesh gave an example of how once when visiting his family outside Banaras he received a call from France late at night. The French tourist whom he had ferried several months before wanted Ramesh to arrange for a sunrise boat ride the next morning for a group of his friends visiting the city. Ramesh immediately called his brother, woke him and informed him which hotel and at what time to meet the group.

Assi Ghat boatmen in 2011 spent an average of Rs 500 a month on calls—five days' wages for an agricultural labourer. Boatmen viewed this spending as a necessary investment in their growing business and expanding customer base. Many of the calls were made to other boatmen, often the workers or drivers who did not own boats or live on the ghats but who operated boats for others throughout the day. Raj Kumar, for example, owned three motor boats and two smaller row-boats; he employed several drivers to operate his fleet, because it was essential for him to remain on the ghat to bid for passengers. The ghat became an office with boatmen constantly on the phone liaising with their workers, monitoring their every move and ensuring they were back on time for the next load of passengers. Like the factory clock and the pocket watch of the nineteenth century, the mobile phone enabled discipline and control as well as independence and individuality.[55]

Priests (*pandas*) in Banaras had long experience of record-keeping as part of a generations-old system in which they and their forebears recorded the names, regions, horoscopes and gifts of clients from all over India. For the semi-literate boatmen, such record-keeping and time management were a major change from the more 'task-oriented' practices to which they were accustomed.[56] With the mobile phone, their lives and working habits were restructured. New disciplines and conventions were created, both among and between boatman and others. This is not to say that capitalism and a market economy were absent until the arrival of mobile phones, and nor is it to conjure up a romantic vision of the boatmen's previous lives as being governed by the 'unchanging rhythms of nature'.[57] The moral economy of the old work structure embodied relations of hierarchy and domination, typified by those that prevailed among boatmen from different ghats.

The mobile phone undermined working systems based on a moral economy of social justice and time-honoured expectations. But it would be misleading to contrast the 'rational', self-maximising behaviour of mobile-phone users with the moral economy of the pre-mobile phone era. Reality was more complex. The boatmen worked in a distinctive occupation with its own codes and practices—a 'living and breathing' moral economy of boating that was constantly debated and contested within the community itself. The mobile phone became another variable in a chain of technological advances that posed challenges and brought opportunities. But the range of services and connections provided by the mobile phone, its ease of use and its very low cost made it a tool like no other for intruding into the lives and livelihoods of very large numbers of people.

On the farm ...

At the beginning of the twenty-first century more than 50 per cent of Indians depended directly on agriculture for their livelihoods. True, the share of agriculture as a proportion of GDP fell to less than 15 per cent by 2010, but this statistic obscured a demographic truth: 70 per cent of India—more than 800 million people—still live in villages.

From when Rajiv Gandhi became prime minister in 1984, governments and private agencies floated programs aiming to enable Information and Communications Technologies (ICTs) not merely to benefit the poor but to attack poverty itself. Results of programs built around personal computers were discouraging. When Rajesh Veeraraghavan and his team worked in the western state of Maharashtra, 'the second-richest in India' in 2008, they found the remains of a computer-based project begun ten years earlier to bring interactive information to villages:

> Most, but not all, of the computers were in working condition ... Cables had apparently been chewed by rats ... The PCs were running Microsoft Windows 95 ... Connectivity was provided by landline telephone dial-up, at a rate of no more than 10kbps [kilobytes per second], and use of the Internet ... was restricted to standard File Transfer Protocol (FTP) to communicate between the [sugar] cooperative's server and the village kiosks.[58]

Veeraraghavan and his colleagues found that the creaky computers performed one function: transferring data from villages to the cooperative headquarters and receiving in return news about when farmers were to be paid for crops.

These findings led them in a similar direction to EKO's—towards devising ways to enable a mobile phone to deliver the services that most people most wanted. Simplicity was the key. They replaced the aged computers with cheap cell phones, one for each of the seven village kiosks where they were conducting the experiment. The idea was that the kiosk attendant would use the mobile phone instead of the computer and that the phone would be faster, more reliable and no longer tied to a room where rats might feast on cables.[59] The system enabled an SMS message, using pre-defined fields, to query a farmer's account with the sugar cooperative and generate an SMS reply providing the farmer's latest entitlements and balances. Farmers liked the system, 'possibly because they perceived mobile phones as a technology that they could understand'. It also gave them quick answers. Their officially measured crop tonnage, which in the past took two weeks to be processed and conveyed to them, was now available by SMS as soon as it was recorded at the cooperative's headquarters. Within eight months, the seven mobile phones in the kiosk had been joined by sixty-one new mobiles purchased by those who wished to join the system and be independent of the kiosk and its attendant.[60]

Veeraraghavan and his colleagues had no illusions about what they had achieved in the seven villages. They had shown that straightforward systems dealing with everyday needs were readily taken up, even by people who were illiterate or semi-literate. Numbers and keypads could be quickly understood and mastered, and voice—the need only to speak—appeared an immensely attractive ingredient. Even in this prosperous region of Maharashtra, 'farmers are mostly illiterate'.[61] Elsewhere in India, where people were poorer and illiteracy even more prevalent, voice messages for rural people also had major appeal. India's low levels of literacy emphasised one of the strong points of the simple mobile phone: the appeal of the human voice. Analysts of the modern telephone have for a hundred years discovered to their surprise that people like to talk to each other. Telephone proponents at the beginning of the twentieth century in North America thought they had to educate people in ways they could use the telephone and therefore why they would want one. However, users quickly discovered the social pleasures of the phone, and those living in rural areas led the way in acquiring them. Until the arrival of radio and the affordable automobile in the 1920s, more rural-dwellers in the USA had phones than their urban counterparts.[62] Ninety years later in India, one of the first surveys of mobiles among north Indian peasants concluded that 'most farmers ... used their phones primarily for social purposes'.[63] They also were among the world's lowest users of text messaging,[64] the result of lower literacy and complications from the fact that India works in 11 different scripts, all of them taught in the schools of their relevant states as the script in which to write one's mother tongue. But texting in the scripts of languages such as Kannada or Oriya was not well supported with mobile-phone software (see Conclusion).

Yet the mobile phone was already altering rural and agricultural practices and held out promise of much greater change. Enthusiasm could be excessive: 'Potentially enormous volumes of use ...', an enthusiast wrote, 'will lead to economic expansion from the bottom up'.[65] But technology does not eliminate political and social structures, though it may modify them. Among peasants and farmers, the wealthier and more skilled could use cell phones to consolidate their positions. An early study estimated that 40 per cent of crop losses could be prevented by early warnings about weather or blight delivered by text messages to cell phones.[66] The larger your crop the more you stood to benefit (or lose), and if you had a larger holding, the chances were you would be among the first to acquire a mobile phone and have sufficient skills to use both voice and text messages. The vision of small cultivators empowered with information about prices that would allow them to move between markets and bargain with middle men had only a few concrete manifestations. A survey found 'some evidence that their bargaining power with traders improved when they were armed with market price information', but the effects were patchy and inconsistent.[67]

Attempts to use mobile phones and digital technology to benefit rural people produced sometimes unexpected results which in turn pointed in directions where society may take technology. IKSL (IFFCO Kisan Sanchar Ltd), for example, was set up in 2007 as a collaboration between one of India's largest fertiliser producers, the cooperative IFFCO, Bharti Airtel, the telecom giant, and Star Global Resources Ltd, a finance company. By the end of 2011, more than a million agriculturalists had bought into the system by buying an IKSL SIM card for their mobile phone. The card—a Bharti Airtel card of course—allowed them to do everything a mobile does, but it also sent them five voice messages a day about agricultural conditions and opportunities and a helpline for specific advice. IFFCO brought 40,000 agricultural cooperatives around India into the program.[68] By the end of 2011, IKSL claimed more than a million active users (and 8 million subscribers) in eighteen of India's twenty-eight states.[69] Though these numbers meant that fewer than two per cent of India's 90 million 'farmer households' used the service regularly,[70] the service was capable of growing—'to scale up ... so that the large rural base is not left [out]', as the IKSL website stated. But it warned: 'Many initiatives fall short in replicability on a large scale'.[71]

The evolution of IKSL showed how mobile phones might be adapted to fulfil people's needs in rural India. For Bharti Airtel, it was worth being involved in enterprises that promised to increase mobile phone use in the countryside. Part of the justification that the founders of IKSL put forward for the project was that in 2006 rural mobile-phone penetration was low: less than two per cent of the population. Within five years, the promoters claimed this had risen to 20 per cent, partly as a result of IKSL.[72] For the fertiliser cooperative, the phone project provided a chance to improve the services to its members; the finance company saw possibilities of finding new ways to mobilise deposits. And the program seemed to make money.[73]

As with EKO, education and promotion were essential. 'We bought and distributed mobile handsets to village leaders', Ranjan Sharma, one of its founders, told a journalist about the early days of the program in 2005. The respected recipients received voice messages about agricultural conditions that they passed on to the farmers.[74] The program began with an old premise—the 'two-step flow'—that messages were influential when they were communicated first to respected local figures who passed them on to their neighbours and associates.[75] The program also underlined the strengths of the mobile phone: it was based on the *voice*—you did not need to be able to read or write—and it was cheap and easy to use. Though the pilot program found it necessary to give phones free to local leaders, by 2007 the take-up of mobiles, even in the countryside, was so strong that the service claimed 800,000 subscribers.[76]

Empowering, ensnaring or just chatting ...?

Experiments with mobile phones ranged from the modest ones of boatmen on the ghats in Banaras to the grand designs of EKO to bring banking to the masses and of IKSL to improve the productivity of Indian agriculture. The magic lay in simplicity and economy. A mobile phone, unlike so much technology of the past two hundred years, was able to be owned and used by the vast majority of Indians. By 2010 the cost of handset fell to around a week's wages for an agricultural labourer, and one rupee (about two cents US) could buy four minutes of talk time. To use a phone required familiarization with a dozen keys on a keypad and memorization of the sequence required for particular purposes. People who could not read and write often had excellent memories. As the boatmen at Banaras showed, a little literacy was enough to make a mobile phone hum with activity. Moreover, the phone prompted improvements in literacy because better literacy brought more ways to exploit the phone.[77]

Being able to bypass reading, writing and print was an immense virtue for a tool for mass use and recalled the enthusiasms of Marshall McLuhan for a 'global village' in which people were liberated from the bondage of print. But the mobile phone did not lead there. Rather, it lured users down a path that made literacy desirable, useful and attainable and initiated them into a world of record-keeping, data-bases and regular exchange with the apparatus of government. A mobile phone user could not avoid the state, but users could now engage with it on more systematic, predictable and equal terms. At the same time, the mobile phone was the most effective device India ever had for locking its people into connections with the state in ways that could be verified and monitored. The state could find and track its citizens as never before.

The examples of cell phones in daily commerce focused on people doing relatively small things. As they went about their mobile-phone-related business, they were also enriching the great capitalist families and corporations that 'owned' slabs of Radio Frequency, built the transmission towers, sold the phones and installed the technology. And government, or terrorists, could shut it all down by threatening the operators, turning off the electricity or blowing up the towers (Chapter 8). Mass mobiles were scarcely a cure-all for poverty or oppression, though in some circumstances, the ability to communicate freely and cheaply tilted balances and changed equations, as the consideration of political organisation, which we explore in the next chapter, suggests.

Illus. 1: Multi-tasking. Ganga River boatman ferries passengers and checks phone. Banaras ghats in background. July 2012. (Photo: Sachidanand Dixit with thanks).

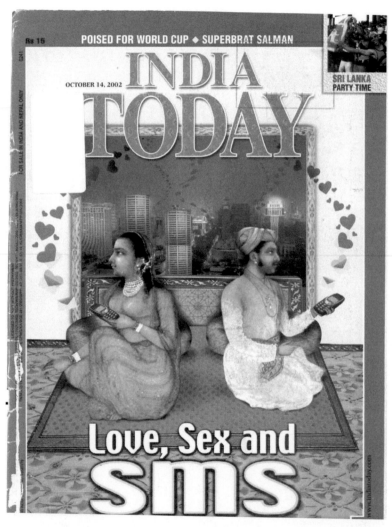

Illus. 2: Love in a time of SMS. *India Today*'s artist gave 17th-century lovers 21st-century devices. (*India Today*, 14 October 2002, with permission).

Illus. 3: Vanishing species. Yellow-painted Public Call Offices (PCOs) were everywhere but are far fewer than 20 years ago. (Photo: Wikipedia, downloaded July 2012).

Illus. 4: *Swamy's Treatise on Telephone Rules*. The 850-page guide to the mysteries of the government telephone monopoly was a profit-maker as late as 1993. (Photo: Robin Jeffrey).

Illus. 5: Duelling telcos. Vodafone and the government provider BSNL fight for attention in Banaras. October 2009. (Photo: Assa Doron).

Illus. 6: Cheeka to the rescue. The mobile-phone dog started her advertising career with Hutch and helped move the brand to Vodafone. Cheeka, like the cell-phone service she represents, is always there when you need her or so the ad would like us to believe. (Ogilvy and Mather and Vodafone, with permission and thanks).

All you need to know

BlackBerry™ - a special mobile phone that lets you do a lot of things you can do on a computer. Like sending and receiving emails, viewing attachments and lots more, even when you are out of the office. **Bluetooth™** - is a wireless technology that lets you transfer images, videos, music etc... to other mobile phones, PDAs and laptops without a cable, for up to 10 metres. **Wireless connectivity** - lets you connect to the internet, computer or a network without wires or an add-on adapter card. **Vodafone mail** - allows you to access emails with attachments on a range of Vodafone mobile phones **3G** - is the third generation of mobile phone standards with technology that enables us to offer you a range of advanced services. The world's most popular mobile telephony technology. **Tri-band and Quad-band** - mobile handsets that **GSM** - you need to roam in North and South America. **Megapixel** - is a term used to measure the quality of pictures a camera takes. The higher the MP number, the clearer your pictures. **MMS** - stands for Multimedia Messaging Service, it lets you send pictures, music and text to any other mobile phone or email address. **Mobile games** - are cool games you can download and play on your mobile phone. You can download mobile games by visiting Vodafone live! on your Vodafone mobile phone. **Memory card** - A card you can insert in your handset to increase storage, so you can save more movies, videos or songs. **Mobile browsing** - it lets you surf sites specially designed for mobile phones. **PDA** - is a Personal Digital Assistant. It's a hand held computer and organiser built into your mobile phone. **Vodafone live!** - our very own online mobile portal that lets you enjoy a new world of news, information and entertainment on your Vodafone mobile phone. **MP3 player** - Software you can install in your phone that lets you listen to your favourite music downloaded from Vodafone live!, wherever you are. **GPRS** - This technology enables you to connect to mobile internet sites via your mobile phone. **Wallpapers** - are images that you can use as backgrounds on your mobile phone. You can download wallpapers by visiting Vodafone live! on your Vodafone mobile phone. **Polyphonic ringtones** - are stereo ringtones that sound more like real musical instruments. You can download Polyphonic ringtones by visiting Vodafone live! on your Vodafone mobile phone. **Real tones** - Actual live recording of music tracks that you can use as ringtones. You can download Real tones by visiting Vodafone live! on Your Vodafone mobile phone. **Roaming** - is a service that lets you make and receive calls on your Vodafone mobile phone even as you travel within the country and across the world. **SMS** - stands for Short Messaging Service, it lets you send text messages to another number without disturbing the other person. **Vodafone alerts** - are updates on a variety of topics like news, sports etc... which flash on your mobile phone from time to time. You can subscribe to alerts you like and stay in the know.

Our guide to plain speaking

Illus. 7: Educating consumers. In one of the many Vodaphone outlets in Banaras, a placard explains to English speakers the wonders of mobile communication—from 'SMS' to '3G' and 'Wifi'. October 2009. (Photo: Assa Doron).

Illus. 8: 'I have the touch', Bollywood superstar Aamir Khan, 'Brand Ambassador' for Samsung, tells buyers. The handwritten notice makes clear to newcomers to capitalism: 'Fixed Prices, No Bargaining'. October 2009. (Photo: Assa Doron).

Illus. 9: The art of retail, Part 1. Ravi's privileged Samsung Mobile Outlet. Banaras, June 2010. (Photo: Assa Doron).

Illus. 10: The art of retail, Part 2. Sales promoters explain features in Ravi's shop. Banaras, June 2010. (Photo: Assa Doron).

Illus. 11: The art of retail, Part 3. Nokia display at Samir's modest shop. Banaras, October 2009. (Photo: Assa Doron).

Illus. 12: Seasonal work. Sumit came to Samir's shop to promote Nokia mobiles prior to the Diwali festival in October 2009. (Photo: Assa Doron).

Illus. 13: 'Make Distance Vanish'. Cheap calls made long-distance romance possible. 50-paise-per-minute (one US cent). October 2009. (Photo: Assa Doron).

Illus. 14: Three towers out of 400,000 across India. North side of Delhi: July 2012. (Photo: Assa Doron).

Illus. 15: Road-side fixer. Lucknow's Hazratganj. June 2010. (Photo: Assa Doron).

Illus. 16: Pavement paraphernalia. Mobile-phone gear includes batteries, chargers, cases and more. Delhi. February 2011. (Photo: Assa Doron).

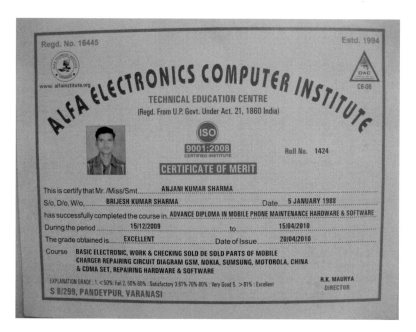

Illus. 17: Mushrooming industry. Diploma from a mobile-phone repair institute. Banaras. February 2011 (Photo: Assa Doron).

Illus. 18: 'Choreography of consumerism'. Nokia's classy service centres may alienate poor customers. Banaras, October 2009. (Photo: Assa Doron).

Illus. 19: 'Nokia laaiif tuuls [life tools]. Valuable information—within your reach'. New Delhi. January 2010. (Photo: Assa Doron).

Illus. 20: Tied to the state. Getting a SIM card requires filling out a form, attaching a photograph and providing personal details. (Photo: Assa Doron).

Illus. 21: Sea cells in south India. Kerala fisherman and their phones gained early fame. (Photo: *The Hindu*, 17 May 2012, with permission and thanks).

Illus. 22: Banking comes to a shop near you. An EKO bank outlet (large white signboard, top centre) and Fast Moving Consumer Goods. New Delhi. February 2011. (Photo: Assa Doron).

Illus. 23: Everybody's doing it. From Airtel's smooth middle classes to nicely posed rickshaw pedallers. Chandigarh. May 2009. (Photo: Ajay Varma, Reuters, with permission).

Illus. 24: Communication technology, old style. Kanshi Ram, organizational genius of the Bahujan Samaj Party, on a cycle *yaatra*. (Photo: http://www.ambedkartimes.com/sahib_kanshi_ram.htm. Photographer and original place of publication unknown).

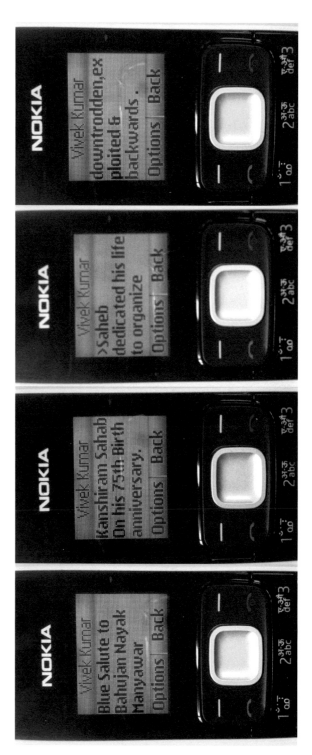

Illus. 25: SMSing to the faithful. Message to Bahujan Samaj Party workers calling on them to celebrate the 75th birthday of their late founder, Kanshi Ram, in 2009. Blue is the party colour—thus 'Blue Salute'. (Photo: Robin Jeffrey, June 2010).

Illus. 26: Communication technology, Mark II. Akhilesh Yadav takes up Kanshi Ram's bicycle and adds a mobile phone for the 2012 election campaign in Uttar Pradesh. (*India Today*, 5 March 2012, with permission).

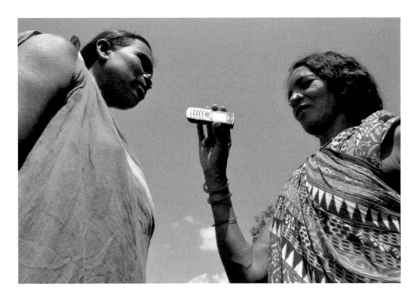

Illus. 27: Communication technology, Mark III. Tribal women send news items to CGNet Swara. (Photo: Purusottam Singh Thakur and CGNet Swara, with permission and thanks).

Illus. 28: Phoning or broadcasting? Tribal woman sends message to SGNet Swara. (Photo: Purusottam Singh Thakur and CGNet Swara, with permission and thanks).

Illus. 29: Buying vegetables? Unsettling society? Mobile phones force families to make choices. Why didn't she ring her vegetable seller and ask him to deliver? Banaras. October 2009. (Photo: Assa Doron).

Illus. 30: Who owns the mobile? India's leading mobile-phone magazine calls itself *My Mobile* and puts phone-wielding women on its cover. Elsewhere, families agonized about whether women should have phones. (*My Mobile*, Hindi edition, September 2011, with permission and thanks).

Illus. 31: A woman with a phone excited some people. Bhojpuri video clip, *Mobile Wali: Woman with a Mobile Phone*. Singers: Manoj Tiwari, 'Mridul' and Trishna. Publisher: WAVE VCD. (Accessed from YouTube on 23 July 2012. Screen dump by Paul Brugman, Australian National University).

Illus. 32: For some, a phone in a woman's hand was a disturbing accessory. *Mobile Wali: Woman with a Mobile Phone*. Singers: Manoj Tiwari, 'Mridul' and Trishna. Publisher: WAVE VCD. (YouTube, 23 July 2012. Screen dump by Paul Brugman, Australian National University).

Illus. 33: 'The Washerwoman with the Mobile Phone'. Cover of DVD entitled *Mobile Wali Dhobinaya*. Singers: Dinesh Lal Gaundh and Noorjehan. Publisher: GANGA, VCD. (Photo: Paul Brugman, Australian National University).

Illus. 34: Youth market. 'Adult film star' Sunny Leone became brand ambassador for Chaze mobiles. Chaze chased younger buyers with the offer of cheap multimedia phones. (*My Mobile*, 12 July 2012, with permission and thanks).

Illus. 35: Terror and an amulet. A burned victim of the Mumbai bombings of 13 July 2011 clutches a connection to aid and succour—his cell phone. (*Outlook*, 25 July 2011, with permission and thanks).

Illus. 36: Brain tumours, heart attacks, cancer and impotence are some of the things mobile-phone radiation can do to you, according to this advertisement on a men's toilet in New Delhi in July 2012. Prabhatam, the advertiser, offered 'radiation-safe mobile solutions'. (Photo: Assa Doron).

Illus. 37: SIM card throwaways. Discarded telecom goods pose growing challenges. The tiny SIM-circuit portion of these cards has been popped out and inserted in a phone. (Photo: Toxic Link, with permission and thanks).

Illus. 38: This adult-sized mound of waste was being picked over by cottage-industry waste-recyclers. Outskirts of New Delhi, 2012. (Photo: Toxic Link, with permission and thanks).

6

FOR POLITICS

If a Sudra ... listens in on a Vedic recitation, his ears shall be filled with molten tin ...
Dharmasutra. *The Law Codes of Ancient India*, Patrick Olivelle ed.

In an old India, caste and class determined how much freedom a person had to travel and communicate. Material conditions also imposed restrictions: roads were poor; conveyances were limited and costly; and letters required literacy, writing materials and messengers. More important, low-status villagers, such as agricultural labourers, were prevented from moving about without the consent of their superiors, and religious texts decreed that low-caste people should be punished for even hearing, much less repeating, certain religious incantations. The constitution of independent India in 1950 aimed to end such distinctions and declared all citizens equal; but the success of legislated egalitarianism was spotty, and immense class differences and prejudices remained. It was only in the first decade of the twenty-first century that cheap mobile phones began to overcome the remnants of these constraints. In the hands of committed political workers, honest officials or outraged citizens, the phone became a simple tool for organising politics, disseminating news and alerting sympathetic authorities to failures of law and government. The cell phone gave people on the margins chances to communicate that were unimaginable to their forebears.

'Smart mobs' in the world

Journalists and scholars quickly recognised the ability of individualised, portable communications to gather crowds. Vicente Rafael's article, 'The Cell Phone and the Crowd', made the Filipinos who streamed into the streets of Manila to demonstrate against President Joseph Estrada in January 2001 as prominent as the Kerala fishermen whom the economist Robert Jensen and others popularised (Chapter 5).[1] 'At its most utopian', Rafael wrote, 'the fetish of communication suggested the possibility of dissolving, however provisionally, existing class divisions'.[2] No such thing of course happened. The president was brought down; the crowds soon dissolved; and the class divisions of the Philippines did not disappear. Clear-headed analysts like Howard Rheingold, drawing on the writing of Rafael and the experience of the Philippines, used the term 'smart mobs' as the title of his remarkable book *Smart Mobs: the Next Social Revolution*, published in 2002.[3] 'Smart mobs', Rheingold wrote, 'consist of people who are able to act in concert even if they don't know each other'. He saw immense potential in the independent communication that mobile phones made possible. 'Groups of people using these tools will gain new forms of social power, new ways to organize their interactions and exchanges …'[4] Rheingold reflected on how cell phones fostered demonstrations against the World Trade Organization in Seattle in 1999 and how students, using simple digital video, recorded violence at a protest in Toronto in 2000.[5]

Two other global events became part of the literature: the South Korean presidential elections of 2002 and the Spanish general elections of 2004. In Korea, the outsider for the presidency, Roh Moo-Hyun (1946–2009), won a surprising victory, even though he was disparaged in mainstream media and had his website shut down for most of the campaign. A large measure of credit for the victory appeared to lie in his ability to reach young voters, recent owners of mobile phones. In the 1997 presidential elections, only a third of young people voted; in 2002, 'mobilized through mobile messages', 60 per cent of young voters turned out to vote for Roh Moo-Hyun, and he prevailed.[6] In Spain, two years later, the

Madrid train bombings that killed more than 190 people happened three days before national elections. The ruling party blamed Basque separatists in the expectation that its strong stand against separatism would win voter approval. The problem was that the charges were quickly revealed to be unconvincing, and connections with al-Qaeda-inspired terror groups became apparent. Though the government had a fairly tight hold over mainstream media, mobile phones bypassed them. Cell phones brought tens of thousands of people into the streets to protest against 'government lies' and transmitted hundreds of thousands of voice and text messages highlighting government deceit and pointing to other sources of information. The government lost the poll. 'This experience in Spain', declared Manuel Castells and his associates, 'coming three years after the flash mob mobilization [in the Philippines] ..., will remain a turning point in the history of political communication'.[7]

Incidents such as these made the unpredictable potential of the mobile phone obvious. It possessed the ability to let large numbers of people generate and exchange information. Such information, unlike a newspaper or a radio bulletin, had no central place of production, and enabled people to communicate rapidly and continually. It also gave them a tool with which to call meetings quickly and send the news widely. Rheingold set considerable store by the fact that individualised communication could bring many people together quickly, create a mob, smart or otherwise, and that this offered ways of challenging the powerful. But he also pointed to dangers: 'surveillance technologies become a threat to liberty as well as dignity when they give one person or group the power to constrain the behaviour of others'.[8] Both he and Castells recognised that the latent power of the technology depended on the circumstances in which it was used. People who knew each other could form a chain of validation, each person conferring on a message her or his implicit endorsement. 'The origin of the message', Castells and associates suggested, was one of the 'critical ingredients of the political power embedded in wireless technology'.[9] In certain conditions, the mobile phone made possible what once would have been inconceivable.

A lively literature has grown up around celebration and debunking of the Internet and the possibilities of digital technology to liberate or enslave. Scholars like Rheingold warned from the outset against 'the rhetoric of the technological sublime' and understood that the technology had 'shadow sides'. Governments could place whole populations under surveillance, and 'terrorists and organized criminals' could find new ways to conquer new worlds.[10] A young Belarusian-American author wrote a book deploring 'cyber-utopianism'—'a naïve belief in the emancipatory nature of online communication'.[11] He quoted at length from Internet enthusiasts whose hopes were belied, and underlined the ways in which bad people and corrupt regimes used technologies to sustain and extend their activities and rule. Misguided faith in digital communications to enable meaningful change could 'turn everyone who uses the Internet in authoritarian states into unwilling prisoners'.[12]

Yet such extravagant critiques did not gel with India's circumstances. They overlooked the fact that in the past some groups were excluded—banned, prevented—from various kinds of communication and information. They were victims of a pervasive exclusion, based on social class and ideas about caste, that few places in the world could equal. From the end of the nineteenth century, such exclusions were challenged, but they remained effective in parts of India in the twenty-first century.

'Smart organisations' in India

The practice of 'untouchability', enforcing discrimination based on caste, was outlawed by Article 17 of India's constitution of 1950. But in practice, caste discrimination continued. Without political will and the power to enforce it, constitutional statements and paper laws about equality were unlikely to mean much in daily life, as activists argued and as ordinary people knew. The lowest-status people were the poorest, those with the lowest levels of literacy, the least chance of education and the greatest dependence on superiors.

B. R. Ambedkar (1891–1956) was an 'Untouchable'. He was also one of the two or three most significant Indian political figures of the twentieth century. A graduate of Columbia University and the London School of Economics, Ambedkar spent his life struggling against vicious discrimination and attempting to organise the dispersed, despised and desperately poor who as 'Untouchables' constituted 15 per cent of independent India. His watchwords, wrote a devotee fifty years after his death, were 'Educate, Organize and Agitate'.[13] All three activities cost money and took time, skill and the capacity to communicate. None of this was easy in Ambedkar's day when nearly all Dalits (a term preferred by 'Untouchables' since the 1970s) were poor, fewer than one in ten were literate and a postcard took days to move from one town to another. In such an environment, one can begin to see how a cheap mobile phone might make a difference. If the Spanish elections of 2004 were a 'turning point' in media history, the Indian election involving Dalits and mobile phones in the north Indian state of Uttar Pradesh in 2007 was a landmark.

Ambedkar's message that political organisation and political power were essential for the uplift of Dalits was absorbed by a remarkable Dalit from Punjab, Kanshi Ram (1934–2006). A relentless organiser and visionary, Kanshi Ram saw government-employed Dalits as the kernel of a Dalit middle class and the spear-point of a movement to capture political power. After fifteen years in government service, in 1971 he became a fulltime activist, harping on the theme that organisation leading to political power must be the goal of poor and low-status Indians.[14] In 1973 he and others founded BAMCEF (All India Backward and Minority Communities Employees' Federation) to mobilise the talents and resources of government servants from low-status backgrounds, particularly Dalits.[15] In the 1980s, Kanshi Ram carried the message of 'organise to win political power' around north India by every economical means available. In 1983, his 40-day 'cycle *yaatra*' took the message of caste-based inequity and injustice across 3,000 kilometres and seven north Indian states. (See Illus. 24). When ridiculed for using such an antiquated method as a bicycle, he foreshadowed the way his successors were to use the cheap mobile phone:

> Trucks, tractors, buses, car and rail are all in the hands of capitalists and those who are holding power ... The very same facilities cannot be available to the oppressed and exploited people ... (The) bicycle is the best weapon for them ... If their two feet are all right they can reach any place to make their presence felt.[16]

When a mobile phone became cheaper than a bicycle, 'oppressed and exploited people' found a new weapon that could reach farther, faster and constantly.

From BAMCEF and other experiences of mobilisation, a political party emerged—the Bahujan Samaj Party (BSP), which began contesting elections in 1984. At first, it was regularly defeated, but in 1989 it contested 372 seats in the Uttar Pradesh (UP) legislature, won 13 seats and took nearly 10 per cent of the vote. Uttar Pradesh is the largest state in India with a population of 200 million people at the 2011 census. If UP were an independent country, it would be the fifth largest in the world, exceeded only by China, the USA, Indonesia and India itself. The BSP's 67 seats in the UP elections of 1993 gave it a pivotal role in coalition governments and enabled Mayawati (b. 1956), a young Dalit woman and Kanshi Ram's discovery, to become the state's Chief Minister for the first time in 1995. She again held the post briefly in 1997 and 2002–03 before the remarkable election victory of the BSP in 2007.

In a country where successful women usually came from powerful high-status families and where low-caste people experienced constant discrimination, something was happening for a Dalit woman to become Chief Minister of the largest state at the young age of 41. An explanation partly lay in small improvements in education and opportunity, but more significantly in relentless political organisation, seeking to build on small social improvements. Kanshi Ram used all means of communication that were within the reach of poor people and took advantage of the reservation of a proportion of government jobs for Dalits, especially in Indian Post and Telegraphs (IP&T). By the 1980s, 15 per cent of lower-order government jobs in India were held by Dalits—in line with their proportion of the population—and Dalits held six per cent of higher-status government jobs. It was estimated that more than two million Dalits worked for the government.[17] During his mobilising travels and cycle *yaatras* in the 1980s, Kanshi Ram could count on telegrams being sent and delivered, telephone calls made and messages transmitted, even though phones were rare and inaccessible. India's network of post offices extended deep into small towns and large villages, and Dalit employees were everywhere. They could usually be relied on to communicate messages about coming speakers and exhibitions and sometimes to organise such events themselves. This was in line with Kanshi Ram's goal of using this incipient middle class of 'educated employees who feel deeply agitated about the miserable existence of their brethren'.[18] Mayawati's father had been an employee of Indian Post and Telegraph.

The UP elections of 2007 brought outright victory (204 seats in the 403-seat assembly) to the BSP, the first time since 1991 that a single party had won a majority in its own right. The result was a surprise: 'no one foresaw such a huge victory for Mayawati', wrote a journalist who had followed the campaign. 'In the end', wrote another, 'Mayawati was right [in having predicted outright victory] and everyone else was wrong'.[19]

The social coalition that brought victory was based on an unlikely partnership between Brahmins, the highest of castes, and Dalits, the very lowest. Together, these two groups accounted for more than 30 per cent of the population of UP. If most of them could be persuaded to vote for the same candidate, such a candidate would have a strong chance of victory in a first-past-the-post, multi-candidate election. But how was such instruction to be imparted convincingly, widely and relentlessly? Mobile phones alone did not produce this election result. But they proved essential. The BSP campaign in 2007 changed the nature of Indian political campaigning by marrying a remarkable grassroots structure with the capacity of the mobile phone to connect, motivate and organise, and to do these things even for groups that in previous times might have had difficulty in moving freely out of their villages. Ideology and technology proved a potent combination. Twenty years of bicycle-powered network-building had created the organisational capacity of the BSP. In the hands of thousands of diligent workers, the mobile phone proved a crucial multiplier. Marginalised people acquired a new and effective tool for struggling and mobilising.

In the UP election of 2007, there were echoes of the Obama primary and national election campaigns in the USA in 2007 and 2008, with similar emphasis on the *combination* of workers and technology 'to help deliver our message person-to-person' because 'trust in ... traditional media sources seemed to be dwindling rapidly'. The 'marrying [of] digital technology and strategy with a strong grassroots campaign' was common to Uttar Pradesh in 2007 and the US in 2007–08.[20]

After independence in 1947, Dalits at election time in UP were often treated as additional voters for the landowners on whose land they lived and worked. Stories of landlords marching 'their' Dalits to the polling booths to vote as instructed were common in parts of UP throughout the first thirty years of independence.[21]

Dalits in many villages were treated like children or animals, and, as we saw in Chapter 1, ancient scripture could be called on to validate their exclusion from information.[22] Dalit movements were restricted, they were forced to ask permission to leave the village or hold marriages and festivals, and they faced beatings or worse if they transgressed. In 1971, Dalit literacy in UP was 10 per cent. In 2001, it had reached 46 per cent.[23] Of UP's 35 million Dalits in 2001, 85 per cent lived in villages; more than a third of those villages had no electricity.[24] At the start of the twenty-first century, Dalits remained predominantly poor, heavily discriminated against, overwhelmingly rural and more than half were illiterate.

When Kanshi Ram suffered his stroke in 2003, the rapid diffusion of cheap cell phones had barely begun. India had about 50 million phones of all kinds, mostly landlines in businesses, government offices and middle class homes. BAMCEF and the BSP could, however, call on their adherents in Indian Post and Telegraphs. A BSP functionary recalled:

> We are Dalits from a village some 60 kilometres [about 40 miles] outside Lucknow, and it was my older brother who introduced me to the party. He was a clerk in the post office, and is now retired, but in those days they used to convey the message through the telephone and tell people when Kanshi Ram was due to arrive in the train-station. This way my brother and others like him were able to organize people from the area to come and meet Kanshi Ram in the railway meeting rooms.... But now the mobile has made it very easy for us to convey our message also at village level. People can recharge for only 10 rupees.[25]

During the same interview in June 2010, another senior functionary had two mobile phones constantly ringing, one of which had a screensaver of Mayawati, the Chief Minister. When asked whether the party paid for their phones or for the calls made on their phones, he replied: 'Nothing is paid by the party; people have and use their own mobiles'. Another person sitting beside him dressed in the khaki uniform of a government servant added that this was precisely why BAMCEF members were required to be educated and have a government job: so that they did not need to rely on the party. 'BAMCEF members are strongly committed to the Dalit cause', he said, 'and their work is carried [on] after working hours, on Sundays and public holidays'. He then switched from English to Hindi to reflect on the past:

FOR POLITICS

> We went from mohalla [neighbourhood] to mohalla, door to door talking to people, even on Sunday and all other holidays...that was a time when people were possessed [*ham diivaane the*] by the ideology; we used to divide areas and go from village to village to organize the cadre and organize night meetings and only then public meetings. Now it is much easier to spread the word.[26]

When lack of ideology and commitment was a constant refrain in Indian politics, the significance of the size and conviction of the BSP cadre was worth emphasising.[27] The BSP did not represent a heroic mobilisation of the downtrodden masses, nor did the BSP government, elected in 2007 and thrown out by the voters in 2012, transform the lives of the oppressed. But in 2007 the BSP had a large body of dedicated workers whose zeal and effectiveness were multiplied many times by their use of the mobile phone.

Mobile phones broke the price barrier in 2003–04 when the cost of a call fell below one rupee for 30 seconds. Mobiles increased in number from 13 million to 33 million in twelve months.[28] By 2005, the cost of a new cell phone was less than Rs 2,000—six weeks' wages for the poorest agricultural labourer, but not an impossible dream for even poor families if a household had three or four earners.[29]

For organizational purposes, what was important was not that everyone owned a phone but that key figures did. For getting people to meetings and for political evangelising, face-to-face was best—as the Obama presidential campaign also believed[30]—though a personal telephone call for a specific purpose was a strong second. More important, the mobile phone enabled meetings and talks to be set up quickly and effectively. And the phone allowed lower-level functionaries to be in regular touch with people higher up in the party and experience the exhilaration of receiving a personal call from a superior and to be asked to provide information or carry out an assignment. The phone connected and energised as a postcard or telegram never could. The number of phones in Uttar Pradesh more than doubled between 2005 and 2007: from 13 million to 31 million.[31] This meant in theory that one in every six people had a phone. Distribution of course was highly skewed towards the urban middle-classes, but even though coverage and penetration in rural areas were patchy, most villages could now be reached by telephone. Those phones were mostly individual mobiles, not locked up in a government

151

office or the house of a wealthy family.[32] And as government servants, often in the Post Office or Communications departments, the Dalit activists who were pillars of Kanshi Ram's organisation were among the first people to be introduced to mobile telephony.

Kanshi Ram died in October 2006, less than a year before the party he founded came to power in its own right in the biggest state in India. In 2005, however, the BSP seemed stalled, capable of winning only about 20 per cent of the vote in Uttar Pradesh, which roughly reflected the Dalit population of the state. Mayawati's biographer wrote that from 2005, 'she sat down with her aides ... to plan' for the coming elections. She increasingly welcomed the advice of Satish Mishra (b. 1952), a Brahmin lawyer who had been advocate-general (legal adviser to the government) of UP and then her legal adviser.[33] The arithmetic was enticing. Dalits constituted about 20 per cent of UP's population; Brahmins were more than 10 per cent. If large numbers of Dalits and Brahmins, coming from different ends of the old social hierarchy, could be induced to vote for the same candidate, it could produce a winning foundation.

At first glance, a Dalit-Brahmin association seemed implausible. In its early years, the BSP had constantly berated high castes. *Tilak taraaju aur talvaar, unko maaro juuti chaar* was a particularly catchy slogan: 'Brahmins, Banias and Rajputs—beat them with shoes'. But Brahmins in UP had been sidelined in political equations, 'orphaned by the changing contours of the two parties [Congress and Bharatiya Janata Party] that had claimed to represent their interests'.[34] From 2005, the BSP, with Satish Mishra as a diligent link with Brahmin groups, campaigned to convince Brahmins and Dalits that a Brahmin-Dalit association made sense and that to vote for the BSP would serve the interests of both. The method of propagating the message, as described by Mishra and journalists, involved dozens of meetings around the state to set up *bhaaichaara samitis*: brotherhood committees reaching down even to the level of individual polling booths. The committees, made up of Brahmins and Dalits, disseminated the message that Brahmin-Dalit unity was desirable because it could bring electoral victory and a government sympathetic to the needs of both groups.

A polling station was based on about 1,000 voters, and with 130 million voters, there were more than 120,000 polling stations in

UP in 2007. The BSP, and the *bhaaichaara samitis* that grew from it, did not create organisations at every polling station; but they established thousands of polling-station-level committees in all 403 constituencies. Dedicated BSP cadres then got to work using the party's six-tier system. At the top of a leadership pyramid, Mayawati and her inner group developed strategy, which was transmitted downwards from state party leaders to divisions, districts, constituencies, sectors and ultimately polling booths. At each level, there were workers willing to devote time and energy—and their mobile phones—to the party. Kanshi Ram had created the framework and recruited faithful workers over years of cycle *yaatras* and touring. 'To this day it is the cycle *yaatris* and the DS-4 cadres who play a role akin to that of Mao's cadres of the Long March', an observer asserted in 2007.[35] Mishra, the Brahmin link man, and his collaborators held scores of meetings, at which local people were inducted into *bhaaichaara samitis* and instructed in how to organise similar meetings themselves. They were also given responsibility for regular liaison with the BSP hierarchy. Mobile phone numbers were exchanged, lists of workers and their phone numbers assembled and thereafter messages were regularly passed up and down the chain by voice and by SMS. Workers received inspirational messages, talking points, tasks, target dates and directions about how to prepare for visits by party leaders. These regular connections with thousands of workers would have been impossible without the mobile phone. (See Illus. 25).

The mobile phone allowed an intricate network to be created, kept fresh and made to do electorally effective work. The numbers involved in the BSP and in the evolving *bhaaichaara samitis* reached tens of thousands. Downward came regular instructions on precise activities that party workers were to undertake at particular times: ensure that all potentially favourable voters in your area were on the rolls; take particular care that all eligible (and potentially favourable) women voters were enrolled; explain why the Brahmin-Dalit alliance was beneficial; remind people of the terrible state of law and order in UP under Mulayam Singh, the incumbent Chief Minister; and emphasise what would be done under a BSP government. BSP slogans were witty and instructional: '*Chad gundon ki chhati pe, button dabaa do haathi pe*' (to get rid of the goondas [the incumbent government], push the elephant button' [the BSP's symbol on the voting machine]).[36] This cell-phone traffic increased as the election campaign began. Such messages came in the form of SMSs in Devanagari and English. A BSP official in Lucknow showed Doron a page in his files headed 'important numbers', which listed the mobile numbers of key figures. Workers at his level maintained regular communication with superiors, especially during the election period when they reported about needs and requirements at sector and booth levels.[37] This top-down direction was constantly informed and refined by bottom-up communication.

How important really was the mobile phone in this system? The question might be put another way: could anything like this have been made to work without the mobile phone? Basic communications in UP were poor, even at the beginning of the twenty-first century. The 'road network [was] one of the lowest in the country' relative both to population and area.[38] Eighty-five per cent of Dalits lived in villages. Wealthier, higher-caste neighbours could sometimes prevent Dalits from moving freely in their own villages and from going out to visit others. Strangers could be prevented from visiting Dalit areas of a village or could find their visit closely scrutinised. To alert and inform hundreds of people needed only a few mobiles. A single call could leap barriers imposed by illiteracy, bad roads, long bus rides, uncertain postal services and hostile neighbours.

The mobile phone helped in another crucial task: explaining to potentially sceptical audiences why Dalit-Brahmin cooperation was a good idea. The success of the *bhaaichaara samitis* depended on telling a convincing story. The BSP's long-standing unpopularity with the mainstream media made the role of personalised explanation especially important. Owned by caste-Hindus and with virtually no Dalits employed in editorial duties, major newspapers and television channels were disdainful of the BSP and often hostile to Dalit-oriented policies.[39] The BSP under Kanshi Ram and Mayawati had reciprocated the disdain and hostility. In 1996, BSP workers roughed up media people who were said to have insulted Kanshi Ram and Mayawati.[40] 'The BSP leader's constant refrain about a biased *manuvadi* [caste-Hindu] media', wrote Mayawati's biographer, 'was an accurate description'.[41] The diffusion of the cell phone counteracted some of these disadvantages. It enriched networks, made widespread frequent communication possible and involved people who would rather speak and listen than read and write.

Satish Mishra recounted how he spent three months from June 2005 touring every district in Uttar Pradesh and covering 23,000 kilometres to form *bhaaichaara samitis* and convince people why they should participate in the committees. It was difficult, he said, but he and his associates explained why age-old prejudices should be put aside and how they could be overcome. Afterwards, these committees were kept in touch and directed by mobile phones. Every member of the new organization, down to the level of the booth committee, had a mobile. The organization did not provide the mobiles; people had them already and were willing to use them for party work. Mobile phones kept workers motivated by reminding them of their tasks and keeping them in regular touch with the party organization.[42]

The ability to converse was crucial. Since the national elections of 2004, politicians had experimented with mass voice messaging; the effects on voters were hard to gauge but mixed at best. Prime Minister Vajpayee's voice messages to hundreds of thousands of voters in the national election campaign of 2004 were judged by many to have become an annoyance. Initially fascinated by the calls and wanting to talk back to the Prime Minister, people soon discovered that the traffic was only one way. On the other hand, a state Chief Minister, Narendra Modi of Gujarat, used the same technique in his state's elections of 2007, and the messages were said to have been relatively well received.[43] However, as the mobile phone became available to 'the masses', the masses expected to be spoken to as individuals. The mobile phone's ability to provide one-to-one conversations was more influential than its mass-mailing capacity.

The BSP organization and its mobile phones helped the party get voters to the polling booths. The elections in 2007 had the lowest overall turnout of voters in Uttar Pradesh since 1985 (about 46 per cent). It was the lowest Dalit turnout (44 per cent) since 1991. However, among the rest of the population (i.e., everyone except Dalits), 2007 marked the lowest turnout (46 per cent) in 30 years and nine elections. When populations lose interest in elections, the party with the strongest organization—the party best able to identify its supporters and get them to the polls—fares best. In 2007 in Uttar Pradesh, that party was the BSP.

The prevalence of mobile phones among officials and activists made these elections remarkably free from intimidation. UP elections had a reputation for bullying and violence, but an increasingly assertive Election Commission of India, armed with fast tip-offs from tens of thousands of mobile phones, imposed unexpected discipline. Polling took a month, carried out in seven phases to allow police to shift from one region to another to supervise voting. The Election Commission strictly enforced guidelines for fair conduct of elections and brought in police from other parts of India to improve impartiality. All this could have happened without mobile phones, but the mobile permitted rapid reporting and response to breaches of the guidelines.[44] All election agents had mobile phones. It needed only one observer and one phone call to draw attention to misbehaviour, even at remote polling stations.

In UP in 2007, this was doubly important for the BSP. First, the ruling Samajwadi Party had a well-deserved strong-arm reputation. In the past, a party with such capacity captured polling booths, stuffed ballot boxes and intimidated voters. Dalits were particularly vulnerable. They could be prevented from voting or coerced into becoming tools of vote-managing thugs. In 2007, however, people with mobile phones—election agents of the various parties, election officials and ordinary citizens—were present at every polling station. The phone numbers of election authorities were widely publicised and citizens encouraged to report irregularities. Many mobiles had a camera. Intimidators ran the risk of being photographed, and the old practices of smashing the camera or seizing the film were less likely to succeed: a photo on a mobile phone could be quickly forwarded to authorities or uploaded onto YouTube. None of these possibilities would have mattered, however, had there not been in place an energetic Election Commission with an effective police force.

Other political parties in 2007 lacked the dedicated structure of the BSP. The mobile phone enriched organisational possibilities, but it did not create them. In the case of the BSP in 2007, a well-drilled army had replaced slingshots with repeating rifles. Other political aspirants saw what the partnership between the mobile phone and the BSP achieved. One observer was P. L. Punia (b. 1945). An Indian Administrative Service (IAS) officer and a Dalit, Punia was principal secretary to Mayawati when she was Chief Minister in 1995, 1997 and 2002. 'Punia knows about the structure and methodology of the BSP', an official told a journalist in 2009.[45]

In the national elections in 2009, Punia won the constituency of Barabanki for the Congress Party with a majority of 167,000 votes.[46] His methods recalled those of the BSP in 2007:

> Before the election ... it is easy to appoint a booth-level worker, but it's very difficult to keep them mobile and keep them activated. And that is what I did through mobile phone, and also through the material I sent off and on continuously before the election. The mobile did work ...[47]

The mobile made it possible to give booth workers tasks, for them to be in touch with the candidate and to feel that they were playing an important role in the campaign. Punia's team gave them constant activities:

> [We told them] what they are supposed to do for the day—that is, the revision of electoral rolls, issue of identity cards, constitution of booth committees: ... that you should have representation of women, you should have representation of Scheduled Castes, you should have representation of Backward Classes, and you should have the representation of the most dominant caste ... So give them [booth committees] some work or the other and not allow them to remain idle.[48]

If the organisation was to be kept ready and useful, the sense of personal one-to-one connection continued afterwards:

> Even now people will approach me [from] ... any village. I ask who was the booth *adhyaksh* [chairman] who was in charge of the booth, what is his mobile number, whether you were associated with me in the last elections, whether you are still in touch with the booth *adhyaksh*. So I will know whether he is my man or not my man. And the booth *adhyaksh* will also get an honour—that 'I am being remembered. I am being consulted on each and every matter if it pertains to my booth'.[49]

Such close contact was only possible through the mobile phone. Given the warmth the phone conversations engendered it was not surprising that party workers 'had their own phones. They were not given by us. [They were] very happy to use their phones'.[50]

In the UP elections of 2007 committed workers of a well-established organisation deployed a new and especially suitable technology. Opponents quickly copied the techniques, and five years later, the BSP government was overwhelmingly defeated at the polls. The victorious Samajwadi Party of the 2012 UP elections was led by Akhilesh Yadav, who at 39 became the youngest state Chief Minister in Indian history. Yadav applied lessons learned since 2007. He cultivated the network of workers and followers established over a lifetime by his father, Mulayam Singh Yadav, and toured the state for weeks, often—recalling Kanshi Ram's 'cycle *yaatras*'—by bicycle. The iconic photograph of his campaign showed him riding a bicycle, mobile phone clasped to his right ear.[51] (See Illus. 26). His party won 224 seats in a house of 403, an even better result than the BSP in 2007; the BSP in 2012 was reduced to 80 seats. Polling was remarkably fair and peaceful, and voter turnout of 60 per cent of the electorate was the highest in Uttar Pradesh history. Every party, it appeared, now exploited mass-based mobile-phone connectivity to chivvy voters to the polls.[52] Mobile phones were only as effective as the people who connected to them. It was not enough simply to collect phone numbers and provide party workers with mobile phones. Workers had to work; they needed to be motivated by beliefs and convictions; and they needed to coordinate to tell their party's story consistently and relentlessly to voters.

Limits, lessons and possibilities

Elsewhere in the world mobile phones and digital technologies produced varying results in electoral politics. In the South African elections of 2009, the Democratic Alliance, the main opposition party, heavily supported by white South Africans, built 'a complex "Obama-esque" campaign website' aimed at computer users. But their attempts to use mobile phones were 'marginal to the campaign', according to Marion Walton and Jonathan Donner.[53] The African National Congress, the ruling party with the largest following among black South Africans, developed a mobile-phone-based site that allowed chat and exchange between users. Although this attracted a 'core group of young ANCYL [African National Congress Youth League] activists and supporters', it fell into disuse after the elections.[54]

The Indian contrast is notable. P. L. Punia used his mobile-phone network after his election to keep in touch with workers and supporters in his Barabanki constituency. For the supporters, there was the reflected glory of being in touch with a Member of Parliament whom they had helped to elect, as well as the possibility of help from their leader in life's struggles. The difference lay in the perceived usefulness of the phone. While the South African election campaign was in progress, the networks built around the phone had some point, but their users were not necessarily party workers with assigned tasks. In contrast, Punia and his party workers had mutual expectations and satisfactions that continued after the campaign. Two-way self-interest kept connections alive in the Indian example; passive connections attenuated in the South African one.

The ease and reach of the mobile phone made it a potent tool for transforming actions that some people intended to be 'private' into actions that became very public. The Rodney King case in Los Angeles in 1991 showed the repercussions of video-recording. An assault on a drunken African-American driver by four Los Angeles police was recorded by a bystander with a clumsy home-video camera. Having been rebuffed by police authorities, the man who made the recording took the videotape to a television station. The broadcast of the tape led to two court cases, riots in which more than fifty people died and the jailing of two police officers.[55] In the twenty-first century, mobile phones gave billions of people the capacity to do what the bystander with the video camera did in Los Angeles in 1991.

From Africa to Russia, simple mobile phones enabled ordinary people to record misdeeds and report them quickly and relatively safely. In Ghana, a coalition of NGOs produced a manual for election monitors in 2004; it contained no mention of mobile phones. However, by the time of the Ghanaian national elections in December 2008, mobile phones had become the monitors' key weapon: 'a big deterrent to politicians because all eyes are watching', according to a representative of the NGOs.[56] In Russia in 2011, 'Smartphones used as election monitor' headlined the *International Herald Tribune* and recounted how 'a small-town mayor' had been videoed offering large sums of money 'to a veterans group in exchange for votes'. The recording was quickly on YouTube, and the mayor, though a member of the ruling party of Vladimir Putin, was fined by a court for breaking election rules.[57]

It is easy to get overexcited by such stories. There was no doubt that mobile phones partly propelled the 'Arab spring' of 2011, aided by innovations like Bambuser—'Share live moments from your mobile or webcam'—a Swedish service that let users stream video from their cameras to the Web in real time. In old-world language, users became 'live TV broadcasters'.[58] The term '*sousveilllance*'—surveillance from below, not above—came into use. Fast transfer of video, photos and SMS messages from an individual to the Web helped defeat the old technique where police or offended parties smashed cameras and exposed film. 'The police thought', according to Mans Adler of Bambuser, speaking of events in Egypt in 2011, 'if we take all the phones, we can control information. But they didn't. The message still got out'.[59]

One does not need the 400 pages of Morozov's *Net Delusion* to spot excess zeal in the enthusiasm for mobile phones as political tools for powerless people. Politics is about how power is distributed and used. Stalin asked how many divisions the Pope had. In the twenty-first century, the question might be modified: how many divisions does a mobile-phone owner have? The answer of course is none, though mobile phone owners could ring for troops, if they had the phone number, if there were troops available and if the commander answered the phone. Phones alone are powerless without responsible third parties willing to act.

Such institutions can develop. CGNet Swara evolved in tribal areas in the centre of India. A journalist, frustrated at trying to report about people living in remote areas and speaking languages that were not part of mainstream India, helped to devise a news service based on simple mobile phones. The Gondi-speaking people of the hilly, isolated parts of the state of Chhattisgarh had no newspapers, radio or television; most had no electricity. 'If you are a tribal', said Shubhranshu Choudhary, a founder of CGNet Swara and a former reporter for the BBC, 'you have no source of any communications'.[60] These tribal people by the late 1990s lived under conditions that made them ready, if accidental, recruits to India's rural insurrections inspired by the ideas of Mao Zedong.[61]

CGNet Swara depended on penetration of mobile phones into remote areas. By the time it began in February 2010, cell phone towers were prevalent enough to cover large swathes even of distant, rugged tracts, and mobile phones themselves were sufficiently cheap and widespread to make them familiar tools, rather than magic wands waved only by the powerful. Indeed, the government of Chhattisgarh was troubled by various cell-phone related ills, such as dozens of illegal towers put up by eager operators, and school teachers and students distracted from their duties by their mobile phones. The towers were threatened with closure, and use of mobile phones was banned in the state's schools.[62]

FOR POLITICS

As with so much that is effective about mobile phones, CGNet Swara was simple. It made local concerns in a local language available to worldwide audiences.[63] (See Illus. 27). Tribal people with a story to tell rang a number and got a message telling them to push '1' on their keypad if they wanted to record an item of news or '2' if they wanted to hear items on the day's bulletin board. The submitted news items were vetted by Gondi-speakers and other journalists, and if validated, put on the server as part of the day's news. An English-language digest went on the website and as a daily email to the computers of subscribers. Digest No. 2736 with eight items, for example, was sent on 29 November 2011. It described a village in Andhra Pradesh in south India where there was no electricity, but the village's six mobile phones were charged once a week when people went to the market town. 'The network is not great', Choudhary said. 'Nagendra [his contact] would call me from a tree top, and I could only leave SMS messages for him to call me because I knew he would get that anytime he was in the network range'.[64] Not surprisingly, there were demands that CGNet Swara be closed for violating India's broadcast laws. Officials and powerful people got offended, Choudhary said, when 'a person from the rural areas reports and it gets translated and the activist in the city takes it to the right place' to get redress.[65] Although only the Indian government was allowed to broadcast news on radio, the mobile phone was ahead of the law: it was not a radio; it was just a phone; yet to those who were exposed or offended, CGNet Swara quacked like a radio. 'Mobile technology', Choudhary argued, 'is better than radio. Mobile is a two-way communication'.[66]

The service not only disseminated news to tribal people as never before—orally and in their own language—it turned travails of daily life in remote places into 'news'. In a celebrated case—tribal India's Rodney King event—people in a village near Bilaspur, 600 kilometres west of Kolkata were frustrated at being abused by a local official who owed them wages for work on government-sponsored projects and refused to sign them up for further employment to which they were entitled. There were four cell phones in the village, and although the village had no electricity, a nearby hill allowed them to pick up enough signal to connect to a mobile network. (See Illus. 28). The next time they waited on the officer and were abused, they quietly recorded the conversation and, safely back in the village, passed it to CGNet Swara where it became an item that quickly caught the attention of government. Mainstream media reported on it, and 'the State government ordered immediate action against the officer'. (The village subsequently received some of its entitlements.)[67]

Shubhranshu Choudhary of CGNet Swara hoped that 'mobiles can truly democratize communications'. Elsewhere in this book, we argued that the mobile phone shared characteristics with The Equalizer, the holster-carried, hand-held repeating revolver that made firearms in the Old West usable by everyone, not just by the strong and the practised. In *Network Power*, David Singh Grewal hints at ways in which mass mobile phones have the potential to enable significant political change.[68] 'Network power', he contended, results from the evolution of 'standards', ways of doing things that become dominant as a result of large numbers of people voluntarily adopting them. It is not that such 'standards' are inevitable and the *only* way; it is simply that, for reasons that are not always clear, very large numbers of people adopt a standard, and the more who do so, the more the power of a standard grows—the more others join or follow. The small snowball rolls into a big, unstoppable one. The 'standard', and of course the people who exercise some control over it, become powerful.

In the past, the capacity to propagate the ideas or the material objects that became 'standards' was confined to elites.[69] But the autonomy provided by mobile phones equipped every holder to become a broadcaster and for the first time gave individuals the chance to build widespread followings and connections. By providing even poor people with the capacity to 'network'—to link up, communicate—the mobile made possible the creation of affiliations beyond the control of the rich and better-connected.

Imagine, as Lee Rainie and Barry Wellman do in *Networked*, that large numbers of people acquired the ability to record and transmit the transactions of their everyday lives—'lifelogging', it's called.[70] The experience of the Baiga people in Chhattisgarh and their abusive local official showed that such recording was not difficult. It could become commonplace—a 'standard'—for poor people dealing with officials or other power-holders to record the exchange, put it on a website and use the pressure of numbers to guarantee that business was done according to rules and laws. Would new standards of public conduct result and relations between people of different statuses change? The old fear of nobilities that underlings would come to know too much and be able to exchange such knowledge widely and freely would be realised. Or perhaps not. Rainie and Wellman also speculate about a future of government surveillance and digital 'double agents that ... report back to the authorities and sell information to corporations'.[71]

Nevertheless, 'India's first mass-mobile phone elections' in Uttar Pradesh in 2007 showed what cheap, autonomous, mobile communications could do. For the BSP and its activists, the cell phone was faster than a bicycle, more powerful than a postcard; it leapt physical barriers, whether imposed by cost, distance or social superiors. It bypassed mainstream media controlled by caste-Hindus, unsympathetic to Dalit causes. It enabled the difficult message of a Brahmin-Dalit alliance to be explained personally, relentlessly and widely. It fostered activists' sense of importance and purpose by enabling person-to-person conversations to instruct, inform, rally and praise. And it ensured a fair, free election by providing rapid access to responsive election officials. In a country as hierarchical as India, such innovations heralded significant disruption to long-standing social and political relations.

7

FOR WOMEN AND HOUSEHOLDS

Now, a ban on mobile phones for unmarried girls.
Times of India, 23 November 2010

Man kills daughter for cell chat with boyfriend.
Times of India, 5 June 2011

Bagpat panchayat issues Taliban-style diktat to women. No love marriages, mobiles or unescorted visits to markets for those up to age of 40.
Hindu, 13 July 2012

Elites and ruling classes tend to live by different rules from poorer and less powerful people. In a notorious example from the English-speaking world, Henry VIII had six wives in a time when an 'average Englishman' would have had one. The powerful change the rules when it suits them, and they are usually first to adopt ideas and technologies that alter prevailing social customs. From washing machines to the contraceptive pill, upper classes have been able to take on new devices because they have the leisure to discover them and the wealth to acquire them.

The cheap mobile phone in contrast has a unique potency: it puts an immensely disruptive device within reach of the poor. It is small and discrete and can be hidden if necessary. And it enables connections and encounters that would previously have been impossible. The effects of its arrival among poorer people in India have been to shake and challenge institutions of authority and both to reinforce and undermine gender roles.

'Man kills daughter for cell chat with boyfriend' read a headline in the *Times of India* in June 2011 reporting a case in a town called Nawada, 90 kilometres south of Patna, the capital of the state of Bihar. The story continued:

> A man allegedly strangled his 20-year-old daughter, doused her body in kerosene and burnt it before burying her. He did this because he found her talking to her boyfriend on a cell phone. The boy has reportedly recorded [the girl's] screams on his cell phone.[1]

Mobile phones were central to these events and their revelation. A police officer told a *Times of India* reporter that

> following an anonymous phone call that a girl was being burned around 10 pm on May 7, he along with forces had gone to the village around 11 pm. They, however, returned empty-handed as they could not locate the place.[2]

The mobile phone did not create the beliefs that generated the violence. Though the young man and woman were of the same caste, her father drew on a long-standing belief that his family was of superior status to that of the man, whose father had been an electrician.[3] In the past, a father could more closely control the people with whom a young daughter fraternised until the time of her marriage. The mobile phone enabled a couple to make initial connections, keep in touch and plan meetings. The sight of the phone in his daughter's hands, and of its being used to connect with her despised suitor, appeared to have sparked ruthless anger.

Similar sex-and-the-phone stories were common. As early as 2002, when India had fewer than ten million mobile phones, *India Today* devoted a cover story to 'Love in the Time of SMS'. It was illustrated with a 'Mughal miniature' showing a courtly ruler and his lover communicating on their mobile phones. (See Illus. 2). The story was an excuse to titillate. 'Some say', the writer enthused, 'that SMS stands for "some more sex"!' Short Messaging Service, she asserted, had 'changed the lifestyles and attitudes of many cell phone users'.[4] The claim at that time was exaggerated, but a cover story in *India Today* signalled growing awareness that the mobile phone, even as an urban middle-class device, was disrupting conventions and opening possibilities.[5] 'It is so private', an enthusiast explained to the *India Today* journalist. 'Privacy' had been rare in most Indian families.

As the cheap mobile phone spread, it raised apprehensions about threats to morality. A caste organisation in Orissa 'banned the use of mobile phones among unmarried girls to prevent them from "going astray"'. The reason, according to a statement by the group, was that 'girls fall in love after they come into contact with boys through mobile phones'.[6] In northern India, 1,700 kilometres away, 'a village panchayat ... banned the use of mobile phones by unmarried girls, a move they feel would prevent them from eloping with their lovers'.[7] These publically proclaimed bans had no force of law, but journalists eagerly picked them up as examples of the tensions that readily-available, easily-concealable phones were introducing into Indian families. In western Uttar Pradesh, a village elder explained a publicised rape in suburbs near Delhi: 'Mobile phones have given a lot of freedom to these girls and that's why they are behaving in a wild manner'.[8]

In taking cheap mobile phones into their daily lives, large numbers of people of poor or modest means found that they affected courtship practices, marriage relations and kinship ties. This story was not one of unqualified celebration. Commenting on experiences in Africa, Anneryan Heatwole accurately described the ambiguities: 'although access to mobile telephones has many benefits for female users, it [is] not a solution to female poverty or gender inequality'.[9] On one hand, such access could generate conflict; on the other, it could provoke desirable change.

In Banaras, Doron spoke with a few boatmen about the roles of wife and husband. Many of them had recently married, and some had young children. They emphasised the 'division of labour' whereby their wives held responsibility for household chores, including cleaning, washing and feeding the children, as well as managing the income the men brought home. They said that men couldn't save money (*aadmi paisa nahin bacha sakta*) and that men just spent it on drinking and gambling. Women, on the other hand, were thrifty by nature, though there were some 'hi-fi' women who were loose with their spending. They mockingly contrasted their own marital relationships with that of their colleague, Dilip, whom they said was controlled by his wife (*vo biibi ka gulaam hai*). The marker of Dilip's loss of control could be seen, they said, in the use of the mobile phone. Carrying her own mobile phone, Dilip's wife ventured outside the house alone to the market or even the riverfront. None of their spouses, they noted, had their own mobiles. Rather, in their homes the mobile phone was kept strictly for family use and to contact the men if the need arose. One of the group suggested that the disruption caused by a wife commandeering the mobile phone was to be expected once families no longer followed the practice of arranged marriage and tradition (*parampara*) and permitted 'love marriages'.

In other times and places, the expanding popularity and social use of the mobile phone's ancestor—the landline telephone—was also controversial and shaped by gender and class. 'Telephone-company managers', wrote Michele Martin of Canada from the 1880s, 'thought that "women's use of men's technology would come to no good end"'.[10] The landline when it arrived in north Atlantic households at the beginning of the twentieth century was viewed with distrust by guardians of social order. 'Some worried', Claude Fischer wrote, 'that the telephone permitted inappropriate or dangerous discussions, such as illicit wooing'.[11] In Pennsylvania, Amish communities saw the phone as a threat to the home and social order, bringing with it the corrupting influences of the 'outside' world.[12]

The mobile phone, a technology so close to the skin and carried wherever one went, could be seen as even more ominous. It enabled people to communicate with unparalleled privacy and independence. How individual families and the world's different cultures adapted the device to daily life fascinated scholars (not to mention marketing divisions and advertising agencies). Scholars used the term 'domestication' to describe the patterns of behaviour that emerged as people acquired objects and used them in their daily lives. Analysing this two-way process of adapting, interpreting and redefining technologies like the TV, microwave and the phone (landline and mobile) became a main concern of the 'domestication' approach in studies with titles like *Consuming Technologies: Media and Information in Domestic Spaces*.[13]

The film classic *The Gods Must Be Crazy* zanily highlighted these processes. The Coca-Cola bottle that fell unexpectedly into the Kalahari Desert dislocated local society. The Bushmen who found it put it to use according to their own interests and perceptions and gave it their own meanings, none of which had anything to do with its being a container for black fizzy drinks. The bottle developed an insidious side. It introduced ownership and property, followed by jealousy and rivalry; relations among individuals and within the community soured. Finally, the Bushman who found the bottle ventured to the end of his known world to throw away the pernicious object. The fate of the Coke bottle dramatised the unpredictable effects of new objects. How much more disruptive was the potential of a new device with the capacity to make any one of its hundreds of millions of owners a broadcaster to wide audiences?

When the Banaras boatmen mocked their colleague for having a wife who wore the mobile phone in the family, they reflected ideas and practices relating to ownership, gender and household economy. Nor were they alone in their apprehensiveness. In West Bengal, the anthropologist Sirpa Tenhunen remarked that the mobile phone had 'increased women's role' in a vital aspect of social life—'marriage negotiations'. Women in the rural areas in which Tenhunen lived now used mobile phones to ring a wide circle of connections in the search for suitable brides and grooms.[14]

The mobile phone tested household dynamics and gender relations; but the complex and fraught relationship between the phone, the community and the individual was not unique to India. Whereas in the industrialised world a mobile phone was usually a private personal item even for young people, in the developing world, especially in low-income families, frequent contests occurred over whether the phone was the personal property of any single member of a household or whether it should move around—be mobile—between family members. Among poor families in Chile around this time, because of the expense associated with the mobile phone, it was considered common property, intended to promote the 'collective good' of the family.[15] In Uganda, 'some ... women were limited in usage of the phone or ... put under escalated control by their partners'.[16] For poor Chinese women migrating to big cities, working and courting relationships were initiated and maintained through mobile phones. The women acquired freedom from the 'restrictive patriarchal conditions in their villages, where they were subordinate in their family and their lives were dictated by housework, farming and after marriage by reproduction'.[17] The Indian context presented similar challenges, though they arose from India's specific historical, economic and social conditions and ideologies.

Doron had the chance to be part of the coming of the mobile phone to the ghats. From the late 1990s, he lived periodically among communities of boat people in Banaras. His informants and friends were mostly men, and he had limited interaction with women, especially young brides. However, studies by women scholars illuminated women's experiences of dealing with the challenges and opportunities presented by the arrival of mobile phones. In the 1990s, phones posed few such dilemmas: the chief accessible telephone was attached to a meter in a yellow-painted Public Call Office (PCO) where a PCO owner supervised calls and took payments.

When mobile phones began to arrive, several questions arose. What was its place in the household, especially in the relationship between a mother-in-law and daughter-in-law? How could a mobile phone be used by young couples in romantic relationships away from the gaze of society? What did the phone mean for people's understanding of themselves as individuals and of their roles as women, men and family members?

Who will guard the mobile?

Doron was invited to live with a boatman and his family. Like many households in the city, the house accommodated two brothers: Vinod, who was in his late twenties, and was married with an infant son; and his unmarried younger brother, Arun. The brothers owned and operated several boats on one of the major ghats. A few days after Doron's arrival, Arun left the city to accompany an old family friend to south India. Soon after his departure Vinod injured his foot seriously. Doron arrived home one day to see him lying in bed being nursed by his wife, and in a matter of days his foot became heavily infected, and Vinod had to be admitted to the local hospital with a high fever.

Despite his absence, Vinod's involvement in the household was maintained throughout his stay in hospital via his and the household's mobile phone. He and his family communicated regularly, enabling him to monitor activities. The mobile phone was carefully managed by his mother, who kept it under her control at all times by tucking it under her sari. This old handset was the only mobile phone in the household, other than those belonging to the two brothers. The majority of households in

Banaras had never owned landlines. Essential calls had been made from PCOs (Chapter 2).

During Vinod's first few days in hospital, his wife spoke with him regularly to inquire about his health. When Doron suggested that his wife accompany them (mother, sisters and Doron) to visit Vinod, his mother quickly replied that there was little reason for her daughter-in-law to leave the house as she was able to speak with Vinod on the household mobile. The mother-in-law/daughter-in-law relationship was strained: the mother-in-law asserted her authority in the house over the newly-arrived daughter-in-law by restricting her movements outside the home as well. Restrictions on women venturing into public spaces were common enough in India, and Banaras was no exception.[18] The restrictions could be traced to cultural ideas based on inside and outside worlds 'aligned with a number of parallel contrasts, including family/not family... safe/unsafe, protected/unprotected, clean/dirty, and private/public'.[19] These, in turn, expressed themselves in gender roles and social structures in the home.[20]

Despite the daughter-in-law's obvious concern for her husband's well-being, she was denied the hospital visit. Restrictions on movement were much more stringent for a daughter-in-law than for Vinod's sisters, who were allowed visits to the hospital. This was partly because the daughter-in-law represented the *izzat* (bearer of honour and reputation) of the household. As it became clear that Vinod would need surgery and would have to stay in hospital for more than two months, his wife was eventually allowed to visit the hospital, accompanied by family members. The daughter-in-law was in a slightly more empowered position than she would have been in past times in that she was able to keep in touch with her husband without having the man's mother as the sole legitimate go-between. The husband had fewer choices to make between his mother and his wife.

The household mobile phone, unlike the ones belonging to the male members of a family which were their own personal handsets, was not considered a 'private' possession of any of the female members. Conversations were conducted in the open, in ear-shot and under the authority of family elders. In very small houses, there could be few spoken secrets. Unlike the household mobile, the men's mobiles had contact lists, songs, video-clips and screensavers, and their handsets were secured with passwords to protect privacy. These were the private mobiles of individual owners and encouraged ideas of 'privacy' and 'ownership'—albeit for men. But gender governed a person's entitlement to privacy and distinguished between different kinds of 'attachment to things'. Literacy, once achieved, could not be taken away from a woman; but she could be barred from having and using her own mobile. 'Keeping in touch' was thus linked with notions of appropriate conduct and possessions for a woman at particular stages in her life. The household mobile in Vinod's home was allotted within the family to maintain a certain social order. Mobile conversations, though an innovation, modified existing social roles and gender ideology only slightly.

The household mobile

Household mobiles among the community with whom Doron lived were often older, basic devices passed on by the men of the house who had purchased new multimedia phones for themselves. For men, the mobile phone was a tool for work, for communication with friends and relatives and for entertainment, while for women the mobile was for 'basic conversations' (*sirf baatchiit karne ke liye*). The woman's phone should be used only to communicate with her natal kin and husband.

There was more than meets the eye, however, in the term 'basic conversation', and the mobile phone began to affect the intensity, and therefore the quality, of accepted social connections. A woman's natal home—her mother and father and brothers and sisters—had long been a source of support for a newly married woman living as junior member of her husband's household. The natal family was often celebrated in songs and proverbs in contrast to the intimidating and much maligned home of the in-laws.[21] By facilitating natal ties, the household mobile phone became a valued possession for newly married women trying to find their way in the new conjugal setting. The phone facilitated more regular connections with her natal family, whether appreciated or deplored in the husband's household; but either way, the phone forced new choices. These possibilities made the household mobile phone a focus of contention which required careful management. If 'unsupervised', the phone was potentially threatening and disruptive. As one informant told Doron:

> My sister-in-law who got married a year back, her husband does not appreciate that she speaks a lot with her sisters, he thinks that whatever problems are created in his family are due to the cell phone—because his wife can speak with her sisters a lot and they are giving her tips and tricks how to handle the mother-in-law. So he banned her from using the cell phone.

A young bride, who may have enjoyed a degree of autonomy in her natal home, had the potential to retain at least a little autonomy in the home of her in-laws through the mobile phone. But elders in her husband's house may have resented and resisted challenges to former practice. A popular Indian soap opera was called *Kyunki Saas Bhi Kabhi Bahu Thi*—'because the mother-in-law was once a daughter-in-law'. What applied in the old days should also apply in the new days.

When Doron expressed surprise about Vinod's wife being barred from leaving the home, even to go to the hospital, friends of Vinod explained that a young daughter-in-law must be particularly chaste: a *pakki bahu*. Other attributes of the 'proper demure daughter-in-law' included wearing a *ghuunghat* (veil) and acting modestly, all of which contrasted with acceptable conduct in her own natal home (*maika*), where she might walk uncovered and be chatty and friendly with members of the household.

In Vinod's home, maintaining control over the household mobile was important. The men purchased the recharge cards.[22] Even when absent, men still exercised their dominant position as household managers and breadwinners. Though weak, bedridden and in hospital, Vinod was constantly on the phone, managing his boating business and his home.

In some instances, lower-class women did have their own personal mobile. Most of these women worked outside the home as domestic workers, school peons or sellers of goods at family-owned stalls. They used the phone to communicate with husbands or for arrangements connected with the business. All of these women were well-established in their husband's home, and some were already mothers-in-law. (See Illus. 29).

Doron arranged to interview one young woman who had her own mobile. As he entered her in-laws' one-bedroom house, he could feel tension. The mother-in-law was seated on the bed behind, carefully listening to the conversation. Throughout the discussion the young woman glanced repeatedly in the direction of her mother-in-law. She said that her husband had recently bought her the phone, a basic Nokia model. Almost apologetically, she explained its importance. With her mobile phone, she, her family and her in-laws felt more comfortable when she travelled long distances to her natal home. Everyone knew that she could be contacted at any time and would be met immediately upon arrival. Thus the inside/outside distinctions were maintained, and the mobile phone was perceived as a tool to manage risk: the risk of a woman venturing into the dangerous and disordered outside world. But part of the danger associated with the outside world was that women left unsupervised were not only vulnerable but also threatening. Though the cell phone defended against danger, it also introduced danger by enabling the outside world to penetrate the 'inside'.

Neither family relations nor inside/outside distinctions were static. Women underwent considerable shifts during their lifetimes, as the above examples demonstrate. Unlike a young daughter-in-law, whose subordinate position in the household was designed to preserve the integrity and honour of the patrilineal group, a woman with school-going children was perceived as less threatening and was allowed considerably more mobility. Her right to use the mobile phone indicated where a woman stood in the hierarchy of family relations at different stages of her life. For the male elders of a village an hour's drive from New Delhi, the magic age in 2012 was forty: they tried to ban women under forty from having mobile phones or going alone to the market.[23]

Experiences varied from region to region and community to community. In the relatively conservative setting of Banaras, concerns about a woman's reticence, modesty and deference seemed to bridge caste and class. When Doron interviewed an educated, middle-class and upper-caste woman living in the same neighbourhood of the city, she explained that prior to her marriage she had owned a mobile phone. During her college years in Allahabad, she had numerous contacts stored in her mobile, including the numbers of friends and classmates, as well as teachers, in case she needed to inquire about an assignment. However, once she arrived at her mother-in-law's house in Banaras, she was required to relinquish her handset. It took her over two years to convince her in-laws (indirectly, by beseeching her husband) to give her a new mobile phone so she could talk to her sisters and family members without needing to use the household mobile. This was fairly common practice in joint-family homes in Banaras, as one man explained:

> All the women who are getting married will not bring their cell phone along with them to their husband's house, and while it's different in different families, until now in my family none of the daughters-in-law brought their cell phones with them. They always leave it at their parents' place and then their parents or sisters use it. Once they arrive at the in-laws place then their husband may buy them a new mobile. A lot of trust is involved. If she will bring her mobile with her into marriage she will receive a lot of calls. Like one of my sisters. She got married and despite previously being a very frequent user, she did not take her cell phone with her. She left it. Later her husband got her a new cell phone. I mean if she would have brought it with her she would continue to receive calls from her friends and then the husband would be doubtful, anyone would be doubtful. After marriage each and everything is exposed to your partner, and if you have a cell phone your husband will certainly go through it…

The mobile phone was viewed as an object of distrust, unless it was monitored by the husband and family. This distrust arose because of the possible flow of 'inside' information to the outside world (e.g., in the form of gossip)—the leaking of family news may threaten the reputation and honour of the household.[24]

With the arrival of a young bride, the family ensured that her previous social network was dismantled. Sarah Lamb called this the 'making and unmaking of persons': a woman should discontinue most of the connections with her previous household when she entered the patrilocal household. Potentially empowering, these connections were a cause for concern for the conjugal family.[25] 'Women's personhood is unique', Lamb wrote, 'in that their ties are disjoined and then remade, while men's ties are extended and enduring'.[26] A woman was expected to cut her ties to her previous networks (symbolised and/or represented by her mobile phone), where she may have had more independence, mobility and freedom, and to re-attach herself to a new kin network and become absorbed into her husband's family. At the same time, her contact with her natal home came under the control of the husband's family. A woman complained to Doron that the mobile phone had reduced the number of times she was allowed to visit her family home, as her in-laws argued that there was no reason to visit if she could call and speak to her relatives. The mobile phone intensified the drama of transition from a young woman to a wife. A device that could keep her connected to her former life raised a question that had immediately to be answered—retain the phone or surrender the phone?—and foreshadowed her relationship with her husband's family in her new home.

Ownership and property

The newly married *bahu* had to appear to adhere to the conduct of an ideal wife, at least until she found enough common ground with her husband to make claims of her own. Mobile phones introduced into millions of even quite poor families a new object of contention, to which a young woman may have had greater access in her natal home than in her husband's. The mobile, moreover, brought with it intrusions from the outside, even as it offered certain benefits of security and convenience. In Doron's experience, contests were usually—though not always (witness Dilip's phone-wielding wife)—settled in favour of the husband's family.

Even when a mobile phone was designated for a young woman's use, it effectively was for 'family' purposes. Raju was in his early twenties and belonged to a socially and economically disadvantaged community (Mallah, or boatman caste) in Banaras. His family earned little from boating, which meant he regularly worked for others, plying their boats as well as tending his family's fields during the hot season. Doron was surprised to see that Raju had not one but two mobile phones. Raju explained that the mobiles had been very useful to him because he recently began working in the catering business. Having a mobile allowed him to communicate with potential clients and arrange his business more efficiently. When Doron asked who else in his family had a mobile phone, he replied that there was one mobile phone used by members of the household and another which he had given to his older sister as a wedding present. She lived with her in-laws in the nearby town of Ramnagar. That mobile, Raju noted, was now mostly commandeered by her husband for his work. Raju explained that his brother-in-law produced small flashlight bulbs, which in the past meant a daily trip to the phone booth to get orders from his clients, which could cost Rs 30 a day in transportation in addition to the cost of calls. By using the mobile, Raju noted, his brother-in-law increased his earnings and saved time. Doron laughed and said that the phone that he bought for his sister was now really for his brother-in-law. Raju nodded in agreement, but added that it was useful for the whole family, as they did not have a landline. When Doron asked if his sister ever got to use the phone, Raju answered that once or twice a week she called her natal home to speak with her family.

Raju's example highlighted how the mobile phone was increasingly becoming part of daily life, a device for managing work and communicating with relatives. But it remained mostly the prerogative of the men. In some cases, it was 'family property': a housebound mobile, rather than being for a woman's personal communication. This was in accord with the giving of dowry to a daughter, which was meant for the daughter's new family.[27] By giving the mobile phone to his sister, Raju reinforced the brother-sister tie in a familiar, long-standing way, central to the familial ethic and kinship system in north India. And while the mobile was integrated into the conjugal family as part of the 'common property', it retained the value and meanings that might be associated with an expression of the enduring and 'mutual responsibilities of brother and sister beyond the sister's marriage'.[28] The mobile phone quickly became one of the gifts expected from a brother to a marrying sister, along the same lines as saris, jewellery and watches. However, unlike saris and jewellery that might be taken away from the bride for the daughters of the household, the mobile phone's utility and visibility meant that it could never be totally appropriated. The bride might be allowed or request—or perhaps even demand?—to use it from time to time. (See Illus. 30).

Although these examples of mobile-phone use appeared to reaffirm established domestic structures and gender relations, this was only part of the story. While the mobile phone was not an outright 'revolutionary' influence changing ways of life and social structures, it did act in incremental ways. It hinted at having long-term effects on one of the most formidable social institutions in Indian society—marriage—and by extension, relations within households and between the sexes.

Romance, marriage and the mobile

Arranged marriage, sanctioned by both sets of parents, remained a deep-seated institution in India in the first decade of the twenty-first century, despite predictions that it would begin to fade away as India 'modernized'.[29] Mobile phones introduced new possibilities for interaction between men and women, particularly for poor people who were semi-literate or could not read or write at all. The mobile phone had remarkable advantages: it was cheap, pervasive and capable of being used independently out of sight of authority. This was where Raju's second mobile phone served as an example.

When Doron asked about his other mobile, Raju began to reveal a more subversive 'mobile practice', associated with a generation of men and women finding ways to circumvent restrictions surrounding marriage and the household. Raju's first phone, he explained, was a Nokia, primarily used for calls in the city of Banaras. His second mobile, a Reliance brand which tied him to Reliance networks, was used for calls outside the city, particularly to communicate with relatives in Allahabad, most of whom also used Reliance. (Calls within the same network were very cheap.) He used his Reliance mobile daily, or more accurately nightly, to call his soon-to-be wife. She lived in Allahabad, 120 km away. Following established practice, the couple had been allowed to meet face-to-face only twice (and in the presence of family) while the marriage arrangements were proceeding. Raju explained that his prospective wife's brother lent her his own mobile phone every night after 10 pm when calls were cheaper. Raju and his bride-to-be were able to talk for many hours into the night. Neither set of parents knew of these illicit phone calls, which had been taking place for several months. The only member of the family who was aware of, and even facilitated, the nightly conversations was her brother.[30]

In this case, the mobile phone complicated and changed but did not destroy long-standing practices. The authority of elders was subverted, but the brother-sister bond acquired a new thread. As for Raju, his mobile was password-protected, and to assure further discretion he used a code name on his phone for his fiancé's number in case anyone in the household or outside got hold of the phone. Raju's descriptions of these conversations were marked by coyness and excitement. He explained how informative, entertaining and enjoyable the conversations were for both him and his bride-to-be. They had learned much about each other's lives, interests, fears and hopes. Doron suspected that for Raju's bride-to-be the conversations were perhaps even more important than for Raju, because she would soon leave her own home and need to learn to live with people she hardly knew.

Such illicit conversations illustrated how lower-income, lower-status people might acquire opportunities to escape the restrictions imposed on couples prior to marriage. Wealthier, higher-status people sometimes had more opportunity, though for them too notions of propriety in relations between unmarried young people were strict. Raju's intended was likely to experience a degree of discomfort in her new status as daughter-in-law; but the familiarity afforded by these conversations outside the purview of authority figures may have alleviated some of her apprehension. The unsanctioned intimacy established via the mobile phone promised to shape her relationship with Raju and influence the dynamics of the home in unanticipated ways.

Like the letter-writing practices discussed by Foucault and Ahearn, the mobile-phone exchanges operated as a 'technology of the self', facilitating the development of 'self' in diverse and often powerful ways.[31] Phone conversations forged knowledge and understanding of one's self and others.[32] The daily exchanges between Raju and his fiancé generated ideas about intimacy, love and companionship, which the couple explored and discussed. The exercise of agency in the courtship—the 'boy and girl' were no longer simply pieces moved by parents on a marriage chessboard—introduced a new dimension into the institution of arranged marriage, with possible implications for the way a couple would conduct themselves within the joint family.

Much of Raju's conversation with his fiancé involved sharing gossip about the foibles and misdemeanours of friends and family. The 'excessive knowledge' with which his fiancé would now arrive in her new home gave her an improper level of intimacy with her husband-to-be and his family. Unless tactfully used, such knowledge might reflect badly on her. However, it might also arm her with knowledge that would keep her out of trouble and enable her more readily to build a place in her new family.

The mobile phone allowed Raju and his fiancé to cross established boundaries of inside/outside. This could generate tensions capable of reshaping not only the nature of the conjugal relationship, but also the dynamics and solidarity of the family. The intimacy forged by the couple prior to marriage might lead them to seek privileges for their relationship at the expense of the joint family; it might also threaten a family's reputation or allow a newcomer strategic insights into the dynamics of a family. Not surprisingly, Raju used his second mobile carefully so as not to reveal his conversations with his future wife. The mobile was a repository of his most private thoughts, desires and activities.

The mobile phone had the potential to transgress and redefine boundaries between private and public. It could alter practices and function as an unregulated gateway for illicit acts and imagery from the 'outside world' to the 'inside' sphere of families. This potential created both aspirations and concerns. The anxieties of heads of families about mobiles were common, and the prospect of an unmarried woman having a mobile phone was a matter of grave concern for some in Banaras. In the everyday lives of the boatmen, the unmarried woman, especially a daughter, was seen as a financial and social burden. The onset of menstruation and the post-pubescent daughter were associated with danger, impurity and risk to family reputation.[33] Little girls sometimes sold flowers and paraphernalia for rituals to pilgrims and tourists on the ghats. Once a daughter reached puberty, however, she was barred from working on the ghats for fear of compromising her own and her family's honour. For many fathers, their primary objective was to marry off their daughters as soon as possible. Once she left her natal home, the father was relieved of the onus of monitoring the young, vulnerable and dangerously fertile person. For these fathers, the mobile phone was an object of distrust that could generate trials and transgressions, as the story of Shiva and Gitika reveals.

Shiva and Gitika met at a wedding in Banaras where they exchanged a few words, and Shiva gave Gitika his mobile number. Gitika soon called Shiva on his mobile from a public phone, and they began speaking and meeting discreetly. Shiva then bought Gitika a mobile phone so that she did not have to venture to a public phone. (See Illus. 13). Eventually, her father found the mobile and confiscated it. According to Shiva, they had done nothing untoward, and following custom, he asked his parents to pay a visit to Gitika's parents to inquire about marriage prospects.

FOR WOMEN AND HOUSEHOLDS

Because of his good disposition and thriving boating business, Shiva felt that the visit to Gitika's parents would make a favourable impression. Gitika's parents were much poorer, but despite her father's appreciation of Shiva as a person of standing, he was unwilling to let his daughter marry outside her caste, which was considered slightly above that of Shiva's Mallah caste. Both Gitika and Shiva were devastated. On Doron's last visit, Shiva and he were talking on the roof of Shiva's house when Shiva's phone rang. It was Gitika calling from the post office phone. Shiva lowered his voice and moved to a corner of the roof. After the conversation, Doron inquired about their relationship. It was over, Shiva said sadly; there was no use pursuing it further.

This example raises several issues. The discreet relations that can be forged through the mobile phone may transgress boundaries of caste and alter courtship practices. Mobile communication enabled new ways of developing ideas about 'the self', about who one was and about how one presented oneself to others. 'Mobile literacy' enabled people to experiment—even with ideas and practices of courtship, 'love' and 'intimacy'. In the past, lower-status youth, unlike middle-class contemporaries, learned the skills, practices and emotions of everyday life from their fairly narrow surroundings and from films. Middle-class contemporaries, on the other hand, read magazines and novels and often experienced the limited autonomy and wider exchanges that went with a college or high school education.[34] The cheap mobile phone extended possibilities for different social interactions. In a story recounted to Jeffrey, a new bride had never spoken to her father-in-law until one day an urgent need to coordinate family movements led her to have to speak to him on the phone. From then on, they periodically had conversations on the mobile phone, though they rarely spoke face-to-face for anything other than household transactions.

The effects of the mobile phone had an insidious side. The mobile's widespread availability brought new experiences and exposure to wider worlds for people whose wealth, education and status would have insulated them from such disruptions in the past. Tensions arose, but change was not always the outcome. Shiva and Gitika did not end up eloping. Rather, Shiva had tried—and failed—to gain the approval of the elders, for a 'love-arranged marriage', a common theme in Bollywood cinema and public culture. This form of courtship sometimes proved acceptable in reconciling 'individual desire and family responsibility'.[35] For Shiva and Gitika, however, the strategy came to nought. Though their mobile phones challenged the values of Gitika's family, the latter prevailed, without the violence and bloodshed that characterised the stories with which this chapter began. The arrival of the mobile phone in a household posed uncertainties and questions. 'The cell phone', concluded Anand Giridharadas, 'gave the young a zone of individual identity, of private space, that they had never known'.[36] But was India unusual in the way the phone affected families, gender relations and the public imagination?

Horst and Miller in their trail-blazing study of the effects of the mobile phone on society in Jamaica were surprised to find that the phone was valued much more for personal social purposes than for commercial, organisational or emergency use. The cell phone, they wrote, 'changes the fundamental conditions of survival for low-income Jamaicans, because it is the instrument of their single most important means of survival—communication with other people'.[37] For women often living on their own (men having migrated to seek work), the mobile reinforced existing practices of keeping in touch with acquaintances to make it easier to get a loan or other support in times of need. The mobile intensified and extended customs that arose 'from a long gestation in Jamaican history and experience'.[38]

The Jamaican example helps to clarify why the Indian experience is both different and remarkable. In India, the privacy of the mobile phone led people not merely to do old things more intensively; it encouraged people to break with custom and do new things. The place of women and the role of gender in India are notably different from most other places in the world—often more defined and more rigid.[39] The mobile phone began to collide with such rigidities.

Daniel Miller highlighted some of the differences that the mobile phone created by contrasting it—surprisingly—with another new phenomenon: Facebook. The popularity of Facebook in Trinidad, Miller guessed, meant that Trinidadians were 'having less illicit or multiple sexual relationships simply because it has become that much harder to keep these from the public gaze'. That was because 'no one ever knows who might be taking a picture of them and posting it on Facebook'. He concluded that in this way 'Facebook and mobile phones work in direct opposition to each other', since the mobile phone seemed to offer a newfound privacy and one-to-one intimacy.[40] In India, the cheap cell phone enabled young couples to talk to each other unknown to disapproving elders, for daughters-in-law to talk to fathers-in-law as they had not done in the past and for such transactions to occur in tens of millions of families almost daily from the early years of the twenty-first century. As these transactions accumulated, like grains of sand on a wind-swept beach, the dunes of social practice began to shift. The shape they would take was unpredictable, but worth watching and studying.

8

FOR 'WRONGDOING'

'WAYWARDNESS' TO TERROR

Dosti Jazz wali se:	make friends with the woman who uses Jazz [the name of a service provider].
Pyar Ufone wali se:	make love with the woman who uses Ufone.
Bat Telenor wali se:	make conversation with the woman using Telenor.
Gift lo Zong wali se:	receive gifts from woman who uses Zong.
Date maro Warid wali se:	go on dates with the girl who uses Warid.
Per:	but....
Shaadi karo baghair mobile wali se:	marry the girl without a mobile!

SMS from a Pakistani mobile user

If mobile phones were as commonplace as footwear, we have to acknowledge that shoes, though they provide mobility, security and status, may also be caked in slime from walking in muck or capped with steel for kicking heads. Users of mobile phones have dark sides. The mobile phone made pornography more widely available in India than ever before—and in high-resolution colour.

Such material ranged from the mildly suggestive to the deeply disturbing.[1] The worst of the latter broke the criminal law or ventured into areas that many argued *should* be governed by criminal law. But the law trailed behind the potential that the mobile phone offered for pornography, fraud and extortion.[2] 'The courts needed to "educate" themselves about how the mobile phone can be used for criminal purposes', a Supreme Court judge lamented in 2012.[3] The mobile phone offered opportunities that no newspaper, telegram or landline could rival to spread scandal, exploit the gullible and coordinate crime, violence, espionage and terror.

The mischief that mobile phones could make was limited only by the imaginations of their users. At an innocent end of the spectrum, they could carry suggestive songs and videos to anyone with the initiative to download them. Such saucy materials provoked outcries that the mobile phone was undermining the values and practices of decent families (Chapter 7). According to such critiques, suggestive songs and video clips led inevitably to pornography, and the mobile phone created a market for pornography that previously was unattainable.

'Wrongdoing' comes in different shapes and sizes and sometimes lies in the eye of the beholder. How best to organise an analysis of the cell phone's contribution to wrongdoing in India? One way would be to approach 'wrongdoing' in terms of social class: from the multi-billion-rupee frauds and tricks of the super-rich to the intimidation and extortion of 'common criminals'. Instead, however, we have chosen to work from the risqué and the naughty to the terrifying and horrendous. Part of the 'equalising' quality of the mobile phone lies in its ability to let poor people become globally recognised—'famous'—by doing terrible things. We begin by focusing on the songs and video clips that scandalised conservatives but were immensely popular and therefore remunerative for those who produced them. Similarly, gossip and scandal, ever practised among human beings, acquired new reach, circulation and permanence. Pornography—explicit and often violent sex—also found a new medium and greatly increased consumption. The mobile phone embellished and refined old-style crimes and allowed the planning, coordination and execution of terror. And to combat all these, the state searched for ways to 'protect' its citizens by monitoring their communications in ways that sometimes could also be construed as criminal.

FOR 'WRONGDOING': 'WAYWARDNESS' TO TERROR

'Waywardness'

'It is the girls who have gone astray', a village elder told a journalist after the rape of a girl near New Delhi in 2012. 'The girls ... are so scantily clad that it's shameful. ... Mobile phones have given a lot of freedom to these girls and that's why they are behaving in a wild manner'.[4] It was a common and unsurprising theme. The autonomy that the phone provided led young people, especially girls, to elude the authority of those who in the past would have controlled and disciplined them. The phone symbolised social disruption.

Advertising campaigns used alluring women to promote mobile phones, and makers of music videos incorporated the apparent liberation bestowed by the mobile phone into songs and dances for DVDs and mobiles. In the first decade of the twenty-first century, it was common in Indian cities to see middle-class women, dressed in Western-style business suits or jeans, using their mobile phones wherever they went. Advertisers tapped into an image of the mobile as an instrument of change. For new, 'liberated' women, the phone was portrayed as a perfect vehicle for gossip (*gupshup*), romance or the promotion of exciting social relations.

Beyond India's cities, and among conservative people in the cities themselves, the mobile phone became a metaphor for changing values and practices related to domesticity, sexuality and morality. In a time of rapid change and disarray, certainties were challenged by ballooning consumerism, relentless migration and unprecedented access to information. The mobile phone embodied the ills of an anxious modernity.

Mobile Wali, a Bhojpuri-language video about a woman with a mobile phone, sung by Manoj Tiwari, was one of many film clips circulating on mobiles in Banaras in 2010.[5] Many songs and videos featured women—popularly known as 'mobile walis'—speaking on their mobile phones to their lovers. Though available in CD/VCD shops and later on YouTube, they were most popular on mobile phones.

The clips featured seductively clad women using mobile phones, dancing in come-hither style and singing lyrics peppered with double meanings.[6] Conservatives dubbed them as soft pornography.[7] This is not to suggest that risqué entertainment did not exist in earlier times. Before it entered the mainstream music market, popular Bhojpuri music was characterised by 'clever phrasing, double entendres, subtle innuendos and suggestive imagery that enabled it to convey taboo sexual acts and desires'.[8] This, however, gave way to what the same critic called the 'raunchy flavour' of Bhojpuri music as it became popular throughout India in VCD/DVD formats and on mobile phones.[9] Bhojpuri music retained its capacity to satirise the 'modern condition' and laugh at the antics of both women and men as they coped with new times and customs.[10]

The *Mobile Wali*[11] clip began with Tiwari (the singer) daydreaming of a woman he met in a bar. It cut to a scene where a glamorous young woman in a halter-neck top, tight jeans and loose hair danced seductively while drinking alcohol and talking on her mobile phone. (See Illus. 31 and 32). This Mobile Wali was depicted as a daring, sexy tease: a woman who defied the norms that usually bind Indian women.[12] She danced, smiled, drank, smoked and wore skimpy clothes—all with a mobile phone in her hand. This was her style, as the chorus said:

> Mobile in [her] hand, she has a smile on her lips.
> She radiates style whenever she moves sideways, forwards, up or down.
> Everyone, including neighbours are dying [from excitement]
> [Because] the babe, having drunk beer... Oh baby, having drunk beer,
> ...The baby (babe) dances chhamak-chhamak-chham.

The following scenes revolved around the woman who made men drool as she strutted around with a mobile glued to her ear. She was both objectified as a *femme fatale* and empowered as someone who could choose from those around her or from others at the end of her phone. The song continued:

> Forever ready to explode with anger [and] swear words on your lips,
> You move the way life moves out of one's body [when one dies].
> The cap worn back to front, dark sunglasses, the cigarette is Gold Flake [a famous Indian brand],
> I'm working at trying [to seduce you], there is still some time to go before we get married.

Portrayed as a loose, urban woman, the Mobile Wali broke long-established rules of conduct, partly empowered by her mobile phone. It could lead a village elder to apoplexy.

Mobile Wali Dhobinaya, a video-clip by Dinesh Lal, was similar.[13] (See Illus. 33). In the first song, entitled 'A Mobile in the hands that normally clutch a rolling pin, My, My!', a low-caste young washerwoman in a white sari walked joyfully in the fields near her village with her mobile phone. (This said a lot about expectations of network coverage in rural India in 2009.) Infatuated with her white mobile, she talked to an anonymous person and kissed her cell phone. The scene of the young maiden dancing in the meadows, uninhibited and daydreaming of love was common in Indian cinema, and the song used this framework to emphasise the changing times. Women were no longer bound by household tasks, such as making chapattis with a rolling pin, but instead were free to roam the fields and speak to their lovers on the mobile phone. Like the *Mobile Wali* clip of Manoj Tiwari, this video lampooned the men, who were depicted as drooling over this free-spirit, 'the mobile woman'.

We found more than a dozen popular songs at this time that highlighted the possibility of connecting young men and women through the mobile phone.[14] The Mobile Wali was anything but the demure maiden presented to a select group of future in-laws prior to an arranged marriage. Rather, she was flirtatious, uninhibited and confident, challenging established social conduct and 'traditional' values. None of this, of course, was 'pornographic' or contrary to the law.[15] Yet for guardians of old values, the unconstrained freedom enjoyed by the Mobile Wali led morality towards dark, wayward ways.

In the *Mobile Wali* clip, the young woman remained remarkably composed, comfortably entering male-only arenas and male-dominated practices, such as drinking alcohol in a bar and smoking in public spaces, all the while talking on her mobile phone. Only among urban sophisticates could such conduct be imagined. The singer and his rustic male companions went to pieces under her spell. The main male character warned his friends: 'She shoots Cupid's arrows with her eyes'. True to the Bhojpuri genre of satire, the clip ridiculed the lewd, drunken men at the same time as it reminded viewers of the challenges that new attitudes and technologies presented to old values. In the next stanza, the young men come upon a Brahmin priest reciting prayers while fingering his *maala* (prayer beads). Surrounding him, they chant verses to the elephant god Ganesh, and the pandit becomes increasingly confused:

> 'Om Gam Ganapati Gam Gavamahe'. [a prayer offered to Lord Ganesh]

It's good if the breeze keeps blowing like this.
Do glance this way, too, mate, sometimes,
Regardless of whether you call me 'a rotten guy' or 'sinner' [these remarks are addressed to the pandit, daring him to join in]
It's awful—this Kaliyug [age of darkness]—[and] Mangaru agrees.
Even Pandit ji here has begun to comprehend the epidemic of fashion.

The clip vividly illustrated the confrontations with tradition that cheap mobile phones provoked. The panicking priest reminded viewers of the precariousness of religious structures and the frailty of people in authority. In the final scene, the priest succumbed to temptation and joined the men in a dance around the woman who still held her magic wand—her mobile phone.

Invoking the notion of Kaliyug, the final and most troubled of the four eras of cyclical time in Hindu cosmology, drew attention to the social and moral degeneration of the digital age. This critique was reminiscent of mid-nineteenth-century Bengal, where the motif was used in popular media to describe the anxieties of the time.[16] The colonial period in Bengal was unsettled by dislocation and the changing status of men, many of whom moved from familiar villages to alienating towns and cities to take up clerical posts. Concerns about the corrosive effects of such movement on social bonds and gender relations found satirical outlets in literature and art, such as in Kalighat paintings. Much of this Kaliyug critique focused on conjugal relations, particularly images of insubordinate, lazy and luxurious women dominating the emasculated landowner-turned-struggling-clerk. The implication was that even the pure Hindu wife—the last bastion of national identity—had lost her immunity to colonial domination and was increasingly infected by Western values and practices.[17]

The *Mobile Wali Dhobinaya* video clip had similar concerns. The attraction and dangers of the village Mobile Wali betrayed a larger anxiety: that of the 'village' divested of its men who increasingly moved to the cities in search of work. The sari-clad wife roamed alone in the fields, with only a cell phone to communicate with her absent husband. The theme recurred in many video clips where the bemused Bihari migrant labourer arrives in the city. He finds a forbidding place, filled with voluptuous Mobile Walis, riding on scooters and confidently chatting on their mobile phones in public. This time, however, it is the Bihari *bhaiya* (village guy) who was ridiculed, a shadow of his former male-self, depicted as helpless and confused at the sight of these city-women with phones clapped to their ears. The mobile phone thus embodied a number of processes, integrated in the familiar motif of Kaliyug, that denoted how consumerism and technology were breaking down 'traditional' moral and social order.[18]

The Mobile Wali-style clips were relatively innocent. Small-time entrepreneurs used the mobile phone—by now, an object of fascination, envy and dismay—to market short entertainment videos based on lively music, suggestive lyrics and saucy dances. Indian manufacturers of handsets, eager to eat into Nokia's dominance, joined the mobile phone and the risque to advertise their phones. The Lava brand marketed its Lava 10 phone in 2010 with a television commercial in which a supermarket cashier gives customers their change in the form of teabags, a common solution to a shortage of small coins. Then a handsome young man, and his even more handsome Lava 10 mobile and its 'sharp gun-metal edges', come to the checkout. The winsome cashier abandons teabags as change and gives him a packet of condoms. Lava, the tag-line declared, 'separates the men from the boys'.[19] In 2012, Chaze Mobile, manufacturers of ultra-cheap cell phones, hired Sunny Leone, a Canadian citizen of Indian origin and a leading actor in pornographic videos, as their 'brand ambassador' for a new range of multi-featured yet very cheap phones. Gambling on the notoriety of Sunny Leone, the company aimed 'to position its product in an extremely cluttered low-end handsets market'.[20] (See Illus. 34).

Pornography

The technology also made possible the easy and discreet delivery of pornography to more people for less cost than ever before. 'Hard porn' became widely available on second-generation mobile phones. According to a Banaras journalist, a locally made pornographic film could generate a profit of about Rs.100,000, first from CD/DVD distribution, which was then converted into mobile-phone format.[21] Films were shot in hotels and apartments. The actors came from the sex-work industry (mostly from neighbouring cities, rather than the locale in which the clip was likely to circulate) or from what was popularly known as 'Music Orchestra Groups'.[22] Salaries for actors ranged from Rs 10,000–

15,000.²³ The producers and directors were local, and the films were distributed in local markets. In Banaras, they could be found in the largest single commodity market—Daal Mandi—alongside an array of pirated goods (Chapter 4), or could be loaded onto a mobile phone from a download kiosk or transferred from the phone of a generous friend by Bluetooth technology. Because the technology was digital, quality did not deteriorate with copying.

Such easy availability and high-quality colour were new. Until the 1980s, Indian printing technology and paper quality were antiquated and poor. At bus stands in north India in the 1960s and 1970s, a vendor might stealthily produce smudgy black-and-white booklets, printed on newsprint and with pictures that only desperate youths found erotic. Indian pornography in those days required outstanding imaginative powers, because the print quality was to *Playboy* what boiled cabbage was to *sambar*. And to find even poor quality such as this required diligence. In 1974, the editor Vinod Mehta researched *Playboy* and *Penthouse* in Mumbai by renting copies 'in dirty brown envelopes ... The hire charges were by the hour'.²⁴ Akshay Sawai of *Open Magazine* pronounced pornography to be a great leveller that found its natural ally in new technologies:

> It just shows how technology, mainly the internet with its ever increasing download speed and easy access, has made all types of Indian males, irrespective of upbringing and age, regular and even compulsive seekers of pornographic thrills. Earlier generations of Indian men did not have such anonymous short cuts to smut. Poor guys only had magazines or video tapes, risky to buy and store. They lived in large families.²⁵

Men recalled that pornography in their childhood had to be sampled discreetly: the police conducted raids and seized printed books from newspaper stalls.²⁶ The booklets contained short stories with grainy, but sexually explicit, photographs. Much of it came from overseas; very little was in any Indian language. Another popular item was the 'advice booklet' which could be discreetly purchased in major railway stations and newspaper stalls.²⁷

Pornographic films were aired in semi-secret cinema screenings in the form of 'cut pieces' or 'bits'. Brief, sexually explicit segments were inserted into the middle of cheaply produced action and horror films, thus camouflaging them from authorities, who usually chose not to probe too deeply.²⁸ In Banaras these films were still screened in 2011 in a small number of lower-grade cinemas that were going out of business because, an informant concluded, people preferred to view pornography privately, on their mobile phones. Mobiles had major advantages over DVDs. To view a DVD required equipment (a player and a television set), privacy (hard to come by in a joint family in cramped quarters) and regular electric supply. A mobile phone needed none of these. In this sense, the mobile phone was classless: it brought porn to the poor.

Downloading services mushroomed to cater to the multitude of people who did not have computers. In Uttar Pradesh, Doron discovered that the flow of sacred images in the form of screen-savers from one mobile phone to another was matched only by the dissemination of Bollywood stars, pro-wrestling champions and pornographic clips. Tiny downloading stations sprang up around Banaras. Many of the boatmen Doron knew loaded their phone's memory card with images, songs and film-clips for a fee of Rs 30–50 per gigabyte of memory. The process took five or ten minutes. The files were often classified under labels such as 'gods', 'heroes and heroines', 'audio' and 'video'. Once purchased from the downloader, these items could be freely exchanged by Bluetooth among one's friends.

Doron struck up a conversation with a downloading-wala in Delhi's Ghaffar Market who offered him a one-Gig package of material for Rs 80. He was then asked if he would like 'blue films' added to the images and music that were downloaded to his cell phone. A conversation ensued about foreign-versus-Indian 'blue films' (*videshi* or *deshi*?). Access to foreign pornography via the internet was easy, and the domestic industry appeared to have expanded to meet the demands of millions of new consumers. According to some informants, the dominance of south Indian pornography, previously the main Indian source, had waned.[29] With the help of new media and webcam technology, pornographic clips were filmed all over India for the mobile phone and CD/DVD market. Production and distribution costs were minimal, and for the more professionally produced clips, the main expense was the actors' fees. The ready availability of the technology also enabled malicious, secretive ways of cutting costs. Study of those methods led deeper into the dark side of the cell phone.[30]

The built-in camera proved to be one of the most popular features of the mobile in India. Even in the 1990s, a visit to a photographic studio was a major event in the lives of poor and lower middle-class people, who would have been unlikely to own a camera. Black-and-white studio photos of solemn families in their best clothes were proudly displayed in huts, shanties and the cramped quarters of urban wage-earners. The camera-in-the-phone changed that, and by 2010, large numbers of people carried a photo library of family, friends and important occasions on their mobile. The technology allowed images to be sent from one phone to another by Bluetooth or by MMS (multi-media messaging service). The phone-camera also made it possible to take pictures easily and quickly in situations where picture-taking was not possible before. Individuals could take pictures of themselves or others in compromising positions—drunk, naked or copulating. As photographic equipment became miniaturised, the pictures could be taken secretly, and then used for blackmail or for transmission to make money or to embarrass the subjects. This was by no means an Indian phenomenon. 'Sexting' passed into the English language as a term to describe the transmission, often by teenagers, of sexually explicit text or pictures of themselves or others.[31]

Across the globe, multimedia phones generated sporadic panics over threats to the moral fabric of society.[32] In India, the landmark event was the so-called Delhi Public School (DPS) scandal of 2004. A correspondent of the *Los Angeles Times* explained:

> A 17-year-old student at the prestigious Delhi Public School used his cell phone to shoot a clip of what police discreetly called an intimate moment with a classmate. It ended up drawing peeks for just under $3 on Baazee.com, India's biggest internet auction site ...[33]

'Now there's sex and the cell phone', wrote an Indian journalist proclaiming the start of a new era.[34] The clip was distributed by MMS and found its way to downloading-walas who gratefully added it to their line-up of salacious material. The incident crystallised for many people the collision 'between the country's traditional values and modern technology'.[35] It partly inspired two feature films, *Dev D* in 2008 and the profoundly disturbing *Love Sex aur Dhoka* [deceit] in 2010. Both films embedded the term 'MMS'—multi-media message service—firmly in Indian languages.[36]

This was the first of many MMS scandals in which the filming was done secretly by cameras in mobile phones.[37] Outrage fol-

FOR 'WRONGDOING': 'WAYWARDNESS' TO TERROR

lowed. Schools and colleges tried to ban mobile phones, and authorities around the country responded to 'the menace' in various ways.[38] 'Reports are trickling in from everywhere', reported Siddharth Srivastava in 2005:

> outside discos, pubs, bus-stops, pavements, colleges. Some policemen have been brazen enough to catch anybody with a cell phone, which is quite easy as there are more than 50 million users in India, and ask to be shown all files, hidden or not, even though this might be beyond the law itself.

Srivastava explained that the police were being especially zealous because of a recent incident. A technically skilled pornographer, somewhere in the world, had taken the face of a popular Indian starlet and transferred it to the body of a pornographic-video star, 'uploaded it into a cell phone and in a blink the MMS was everywhere'. The outraged Indian actress rang everyone she knew to try to eradicate the clip and find the culprit—apparently with no success. 'The market for MMS and porn CDs', Srivastava concluded, 'only seems to grow'.[39]

Six years later, with tens of millions more mobile subscribers, the scale of the homemade pornography industry had increased enormously. A scholar based in Bengaluru in south India wrote:

> There is recognition of cyberspace as producing infinitely uncontrollable conditions of pornography, which can enable, very inexpensively, a huge part of the population to become pornographers in different roles—as produces, as performers, as consumers.[40]

What damage would such pervasiveness do to society? How was such widespread licentiousness to be dealt with? India grappled with these questions, just as other countries had. 'The clean distinctions between communications media and broadcast media' dissolved.[41] Nisha Susan, a journalist with *Tehelka* magazine, noted the vast expansion of opportunities and focused on what she called 'India homemades'. The internet was 'awash' with them:

> Two genres of Indian homemades are considered blue-chip. 1) Fake/real sex tapes of Bollywood or political celebrities. 2) The 'leaked' tape from an educational institution. IIT, JNU, Noida B-School—these labels are instant narratives for viewers to hang their fantasies on.

'The phone camera', she concluded, 'is this decade's sex tweak'.[42] The consequences of naiveté or nastiness could be dire. A couple in Kerala were reported to have committed suicide after clips of them kissing were posted on the internet and circulated via MMS. 'Misuse of camera-enabled mobile phones' was reported to be 'on the rise' in Kerala.[43] In Mumbai, louts used a mobile phone to record a kidnap and rape, subsequently threatening the victim with the release of the clips via MMS if she approached the police.[44] Sometimes the capture of such MMS clips served as evidence of crimes, as in the case of a gang-rape near Mumbai in 2011.[45]

Authorities tried to impose new regulations on content and to implement measures to prevent viewing and protect privacy. However, the immense popularity of mobile phones, and the proliferation of data, made attempts to monitor the traffic both expensive and fallible. *India Today* underlined the difficulties faced by monitors:

> And now there are even porn applications. Imagine a 'pocket' girlfriend or boyfriend, who can strip, talk dirty, make sexual noises. 'These are some of the "apps" that can be downloaded on smartphones', says Pranesh Prakash, programme manager with Bangalore-based think-tank Centre for Internet and Society. 'App download data shows the popularity of sex-themed apps on smartphones, apart from the adults-only stores', he says. Age restrictions for applications? Mostly a pop-up asking if one is over 17. With over 50 per cent of all Internet users in the country accessing the web via mobile phones already, as estimated by TRAI, smartphones are the future of anytime-anywhere porn.[46]

Pornography could be made available everywhere—from *kaccha* houses to penthouses. And though police and morality crusaders aimed mostly at the poor, the powerful too were vulnerable to the seductive properties of the cell phone. An incident in the Karnataka state legislature came to be dubbed as 'Porngate'. Two Members of the Legislative Assembly (MLAs) were caught viewing what were said to be pornographic clips on a mobile phone in the assembly while a debate was going on. The legislators belonged to the Hindu right-wing Bharatiya Janata Party (BJP), constant advocates of censorship in the name of preserving morality and Hindu values.[47]

The mobile phone was *not* the old landline that had slipped into daily life in the USA as unnoticed as 'food canning, refrigeration and sewage treatment' and become simply 'mundane'.[48] The

FOR 'WRONGDOING': 'WAYWARDNESS' TO TERROR

mobile phone, argued Clay Shirky, meant that 'the old habit of treating communications tools like the phone differently from broadcast tools like television no longer makes sense'. The potential to record and to broadcast, at one time limited to those who controlled presses and transmitters, was now available to the majority of people, even the poor.

Crime

As well as improved ways of conducting old-fashioned crimes, the mobile phone introduced new categories of criminal activity. The most obvious, of course, related to mobile phones themselves. Small, desirable and easily concealed, they were stolen every day. By 2011, the phone was so common that a survey by a security company claimed that '53 per cent of the adults in India' had lost their phones or had them stolen.[49] The Telecom Regulatory Authority of India (TRAI) had issued a 'Preliminary Consultation Paper on Mobile Phone Theft' as early as January 2004 when the country had fewer than 35 million mobile-phone subscribers. The law, however, trailed the technology.[50]

To list the crimes a mobile phone could initiate was to underline the fact that a policeman's lot in the digital age had become a technical one. The list began with simple theft. Thieves soon discovered, however, that having a stolen phone meant they could be located and caught. Enterprising thieves learned about IMEI (International Mobile Equipment Identity) numbers and how to change them. The IMEI number is the phone's fingerprint, its unique identity, usually inscribed in the battery compartment. Thieves—and those hiring them—were able to alter IMEI numbers electronically to give a phone a new and respectable 'fingerprint'. From 2008, such tampering was made a criminal offence, but being relatively easy to do, it continued.[51] Moreover, until 2009 it was possible to import and use 'China mobiles' that had no IMEI numbers. The owners of these cheap phones, which were sold in the grey market (Chapter 4), left no electronic fingerprints. In the wake of the Mumbai terrorist attack in November 2008, the Government of India required service providers to record an IMEI number before they connected a phone to their network. This put new limitations on the 'China mobile' trade, which, one website speculated, had 25 million handsets in use around India.[52]

To steal a phone and sell it was old-time law-breaking, little different from stealing a sheep or a silk handkerchief. The mobile phone, however, increased the sheer numbers of people vulnerable to property crime. For many, their mobile phone was the first item they had ever owned that was worth stealing. If the 2011 survey were correct and 53 per cent of Indians had lost their phone or had it stolen, 400 million Indians had experienced the anger, fear and frustration of consumerist crime. It did not threaten lives, but it could break hearts if a hard-earned phone held photos and an address book and provided a regular link with a distant family, as it did for millions of migrant workers.[53]

People resigned themselves to phone theft. One victim described a forlorn attempt to report how his phone had been stolen during his morning bus journey:

> So, I went to the nearest police station and told them my story. They were helpful in that they listened to what I had to say. They asked a few questions about where I lost my phone and after I patiently answered all their questions, they said that I had to complain in a different police station. I was perplexed as to why I had to go to a different station which was very far away when I was already in a police station...[54]

The police, of course, wanted to avoid the paperwork and to keep the crime statistics of their station looking favourable; this victim gave up. The *Hindustan Times* reported ironically that 'the south district of the Delhi police seemed to have performed an impossible feat ... retrieving an allegedly stolen mobile phone'. In New Delhi, the owners of 10,000 stolen mobile phones persevered and registered thefts in the first nine months of 2010, up 15 per cent over the previous period in 2009. The newspaper story reported with a straight face that the police in the West Delhi station area must be doing a splendid job because 'only 22 complaints were received', while in Northeast Delhi, the registered phone-theft figure was 2,800.[55] Some police stations were more successful than others—some in finding phones, others in fending off luckless victims.

Mobile phones enabled more sophisticated and profitable crimes than simple theft. By 2010, 17 per cent of Indian adults were estimated to 'have experienced cyber crime on their mobile phone'.[56] The familiarity of mobile phones lulled users into misplaced trust. They responded to fake calls, gave away personal

FOR 'WRONGDOING': 'WAYWARDNESS' TO TERROR

identification numbers and opened themselves to having their pre-paid credits skimmed off to pay the bills or boost the credits of tricksters.

The mobile phone provided new scope for old crimes like pornography and fraud to find far larger numbers of practitioners and victims. The mobile's democratic qualities, which gave almost everyone a chance to communicate, also exposed almost everyone to the risk of being defrauded. 'The fortune at the bottom of the pyramid' for honest entrepreneurs also beckoned tricksters. 'Beware! Your mobile balance can reach zero', warned a Hindi-language story from a website based in Bihar, one of India's least developed states. The story alerted mobile phone users to a clever fraud that sucked their pre-paid phone time dry. A phone owner would receive a call and hear a girl's 'sweet voice'; the call would be cut off almost at once. Curious recipients would call back, 'and as soon as you call back, your phone balance is emptied'. The thieves appeared to have constructed a system, similar to those used to make money from phone-in-to-vote competitions, by which calls to their number were charged at a very high rate and they pocketed the profits. The crooks targeted areas of the country, such as remote corners of Bihar, where literacy was low and awareness of electronic crime limited. The newspaper story also warned people about the sorts of frauds, well known elsewhere but fresh and new in Bihar, based on promises of lottery wins and prizes once a tax or a fee was paid.[57] This small-time fraud illustrated the power of the mobile telephone: hundreds of millions of users, all of them exposed to new and mysterious ways of being defrauded. If the cell phone brought pornography to the poor, it also brought new ways of taking what little money they had.

Economic crimes were not the only crimes that mobile phones made easier. Cheating, stalking, intimidation, bullying and sexual harassment all came within reach of the common man or woman, both as practitioners and victims. Boys and girls could also join in more easily than ever before. Large-scale examinations, the most common form of assessment in India, invited innovative ways of cheating, which Punjab University countered with 'metal detectors in all of its examination centres' in 2012. 'Using technology like mobile phones and Bluetooth devices', the university concluded, 'has become a common practice during the examinations'.[58] The first All India Post Graduate Medical Entrance examination in 2012 produced an elaborate cheating attempt that ended badly for the perpetrators. It involved document scanners in mobile phones concealed in long-sleeved shirts. The conspirators clandestinely scanned the exam paper and transmitted the image to a 'control centre' where the questions were answered by specialists. Bluetooth devices stitched to the shirt collars of cheating candidates brought the answers to tiny concealed earphones. Six examinees were said each to have paid between Rs 3 and Rs 4 million (US $50,000–$75,000) to receive the answers.[59] In terms of social class, this was elaborate, elite crime on a different scale from simply reselling stolen mobiles.

There was nothing uniquely Indian about using mobile phones to cheat in exams. From Japan to the USA, such things had been happening for years, as had the use of mobiles to stalk and harass for reasons ranging from office rivalry to sexual frustration and exploitation.[60] The advantage for perpetrators, as Ling and Donner wrote, was that they were 'more anonymous'. The bully did not risk being exposed in the playground or at the village well, and victims could be tormented by calls and text messages even in their homes.[61]

The potential of the mobile phone to act as a broadcaster made it dangerous, both as a facilitator and a provocateur of crime. Enterprising kidnappers and fraudsters used mobiles in predictable ways: handy innovations adapted to old trades. But the phone also led the unsuspecting into unanticipated dangers. This was particularly noticeable in matters of romance and domestic discord, as we saw in Chapter 7. A vignette captured the essence of such stories. 'Woman held for killing mom-in-law', the *Times of India* informed readers in July 2011 in a story from Uttar Pradesh. A mother-in-law went missing, her son reported her disappearance to the police and her body was found sixty kilometres from their home. 'When the police checked the call details of the family members'—ten years earlier few police in Uttar Pradesh would have recognised a mobile phone if they had found one—it appeared that the daughter-in-law of the missing woman had been making frequent calls to another man. The plot thickened. 'The call details' of the Other Man's 'cell phone confirmed his presence' near the scene of the crime. As the story unfolded, it appeared that the mother-in-law had discovered the liaison, and her daughter-in-law and the Other Man killed her to silence her. After the Other Man had 'slit her throat', he called his lover on her cell phone 'and told her that "the work has been done"'.[62]

In this example, the murderers used the phone to plot the crime. In other cases, mobile phone records betrayed liaisons, or supposed liaisons, that led family members to kill allegedly wayward sons, daughters or suspected lovers.[63] In a variant of the 'cell phone gives false confidence' theme, a civil servant in Maharashtra was burned to death while taking pictures on his cell phone of the adulteration of expensive oil with subsidised (and therefore cheap) kerosene. He was said to have stumbled across the illegal activity, got out of his car to take pictures for evidence and was set upon and murdered by the angry gang. A later version suggested that the official was demanding bribes, and the gang murdered him over a disagreement about timing and amount. In any case, the mobile phone was central to the challenge the dead official had posed to the gang, whether he threatened them with arrest or quarrelled with them over extortion. The police said the phone was recovered, though what was found on it was not revealed.[64]

Cell phones became one of the first items police looked for at the scene of a crime or confrontation. On one hand, pictures and phone records could reveal the identity of criminals; on the other, if the incident involved the police in an unflattering or compromising way, the evidence on a cell phone was best destroyed before it incriminated them. But the phone had a capacity to transmit messages quickly; destroying one phone did not necessarily mean the evidence itself was gone.

The phone became a lightning rod that attracted a variety of Indian criminal activity—from corporate crime involving phone tapping and industrial espionage through to blackmail, crimes of passion and harassment, kidnapping, petty theft and small-time fraud. Mumbai cinema quickly picked up on the excitement, anxiety and plot possibilities that pervasive mobile phones presented. Films such as *Aamir* (2008) and *A Wednesday* (2009) focused on corruption, kidnapping and terrorism in Mumbai and the way in which mobile media abetted such crimes. For producers and scriptwriters, the mobile phone could be made to stand for 'fear, paranoia, and the uncertainties of life'.[65] In *Aamir*, the protagonist received images of his kidnapped family on the screen of a mobile phone. To free them, Aamir had to plant a bomb in a crowded bus. The mobile phone served as his only connection to the terrorist network holding his family hostage, and through the phone they tracked his movements and issued instructions.

Scandal and surveillance

The phone's capacity to make people sociable—to make it easy to chat and gossip—had been recognised in the days of Alexander Graham Bell and the early landlines. In India, if the cost was small, the poor were no different: they liked to talk too. And the wealthy found themselves talking too much for their own good. The common-placed-ness of the phone made people forget that talking on it was less confidential than mailing a postcard. Private gossip readily became public property.

Great corporations, leading politicians and some of the wealthiest and most sophisticated people in India were involved in the cell-phone industry. Thousands of crores of rupees (billions of dollars) were invested in it in the twenty years after 1995,[66] and private fortunes could be made on the basis of decisions about the legal framework of telecommunications. For the rich and the influential—and those who wished to be—the mobile phone became an item so commonplace that they forgot its latent capacity as a broadcaster. Sophisticates appeared naïve, lulled into a misplaced sense of intimacy and face-to-face exclusiveness, when using mobile phones. The most famous of publicised private conversations were the so-called 'Radia tapes', made public in 2010. Niira Radia, a well-connected public relations consultant, had two of India's most elite companies from the Tata and Mukesh Ambani groups among her clients. During the negotiations to form a new government after the national elections of 2009, Radia lobbied for particular candidates to be appointed ministers in portfolios where her clients had interests. Minister of Communications was such a job, and she lobbied to have A. Raja, a Dalit member of the DMK, which was then the ruling party in Tamil Nadu state, appointed to the post.

There was probably nothing illegal or even unusual in what Radia was doing—eliciting information and cajoling influential people. Unknown to her, however, the conversations were being recorded by the Income Tax department. There was nothing illegal, either, about a government department, provided it had a court order, tapping the conversations of someone it had reason to believe was cheating on tax returns. The recordings, however, were subsequently leaked, almost certainly illegally, to an activist lawyer who submitted them to the Supreme Court in support of public-interest litigation aimed at exposing corruption over the allocation of 2G Radio Frequency spectrum. The recordings were leaked again—no doubt, illegally—to the news magazines, *Outlook* and *Open*. The transcripts and recordings became public in November 2010.[67]

A. Raja resigned as Communications Minister in same month, was arrested in February 2011 and imprisoned without bail while his trial for corruption over the allocation of 2G spectrum in 2007–08 began (Chapter 2).[68] Mobile phones entangled Raja in two ways. First, the corruption charges in which he was embroiled grew out of the struggles of mighty business and political interests to gain rights to Radio Frequency spectrum at bargain-basement fees. Second, the banality of the phone led to the indiscreet conversations that formed the 'Radia tapes' and sank Raja and others deeper into the mire. A leading journalist, not implicated in the recordings, wrote: '104 intercepted telephone conversations … cut through the tiniest cross-section of a rotting cadaver known as the Indian Establishment'.[69]

The banality of the phone led those who used it to forget that the technology did not make it private. The phone could be used to track movements, and conversations could be readily listened to by anyone with simple equipment and modest training. At the same time that the 'Radia tapes' were preoccupying Indian news media, it was estimated between 5,000 and 6,000 phones were being 'tapped on an average across the country' each day by government intelligence agencies.[70] The sanction of the Home Secretary (the top public servant) of the central government or a state government was necessary to allow legal tapping of phones, and it was illegal for private citizens to intercept calls. However, the Department of Revenue Intelligence of the central government estimated that 1,100 'tapping devices' had been imported between 2009 and 2011, but no one was very sure where they had gone. Private detectives and private industry were two likely destinations. When the Home Ministry in 2011 ordered state governments to 'trace and hand over to the Intelligence Bureau' all tapping equipment in their jurisdiction, only two responses came in—from the Delhi police and the police of neighbouring Haryana state.[71] No privately held equipment was surrendered.

Apprehension about the dangers to privacy from new technologies was long-standing and worldwide. Kipling's verse, 'A Code of Morals', written in the 1880s, described the embarrassments a heliograph message could cause when unexpectedly read by the 'most immoral man' about whom a young bride was being warned. But the tale involved a heliograph machine, Morse code and a major-general; unlike the mobile phone, this was not technology for the *aam aadmii* (common man).[72] In China in the twenty-first century, mobile communication was becoming, according to one scholar, 'a "wireless leash" that shop-floor management can use as a nearly complete control and surveillance system over employees'.[73] To critics, 'information technology' looked more like 'the oppressor' than 'the liberator'.[74] It enabled the state to locate citizens, listen to their conversations, read their messages and assemble information about them. The grim future scenario, 'A Walled and Surveilled World', imagined by Rainie and Wellman emphasised surveillance, control, self-censorship and fear.[75]

Freelance phone-tapping could be conducted relatively easily in India. 'Different parts of the interceptor', a reporter was told about monitoring equipment, 'are often imported as separate equipment and subsequently assembled'.[76] Praveen Swami, the well-connected correspondent of the *Hindu*, drew attention to the fact that two Indian companies, ClearTrail in Indore and Shoghi at Shimla,[77] produced sophisticated monitoring equipment. Swami highlighted the way in which such equipment could escape from government control. In Punjab, one out of four units was said to have disappeared, while in Andhra Pradesh authorities 'shut down … passive interception capabilities after [they] accidentally intercepted sensitive conversations between high officials'. In Karnataka, the police 'accidentally intercepted conversations involving a romantic relationship between a leading politician and a movie star'.[78]

Three things combined to make such activity grow. First, it was relatively easy to tap mobile-phone conversation. Second, the vast majority of users treated phone calls as if they were secluded conversations in an empty room, instead of radio broadcasts. Third, the low salaries of many police and intelligence officers meant that secrets and scandal could be bought fairly cheaply. 'When an officer on a salary of Rs 8,000 a month has pretty much unrestricted access to this kind of technology', a senior policeman told Praveen Swami, 'things will go wrong'.[79] Such snooping enabled blackmail, malicious gossip and settling of scores between political and business foes; but such activity usually had repercussions only within the borders of India. Spying, terrorism and violence, however, often had international implications.

Espionage and terror

Throughout the world, mobile phones became tools for espionage, crime and terror. Indeed, mobile phones and electronic surveillance were part of Western crime dramas well before the cell phone began to make a major impact on India. The acclaimed American television series, *The Wire*, began in 2002 with a series built around drug crime and phone surveillance in Baltimore, and its third series focused on cross-border crime where pre-paid mobile phones, dubbed 'burners', were used by drug-trafficking gangs and quickly ditched to escape police surveillance. Terrorist attacks were coordinated and carried out with the assistance of mobile phones from the 1990s. Israeli intelligence in 1996 used a mobile phone to assassinate the Palestine bomb-maker, Ayyash, known as 'the engineer'. Ayyash accepted a bomb-laden mobile phone from a trusted companion. Using the phone detonated the device and killed him.[80] In the Madrid train bombing of 2004, one of the bombs was connected to a mobile phone that functioned as timer, but the alarm, which should have triggered the bomb, failed to go off. Examination of the SIM card and handset led police to the alleged terrorist.[81] (The other side of the mobile phone's role in the Spanish attacks—the rapid dismay of the public at the disingenuousness of the government—is discussed in Chapter 6).[82]

India had foretastes of the temptations and the perils of the technology. During the bloody little Kargil war between Indian and Pakistan in Kashmir in the summer of 1999, Indian intelligence tapped a telephone conversation between Pervez Musharraf, chief of the Pakistan armed forces, who was visiting Beijing, and a senior military colleague in Pakistan. The conversation revealed that the forces that crossed the Line of Control were Pakistani regulars, not freelance 'guerrillas', and that Musharraf had kept his political superiors and air force and naval colleagues ignorant of the plans for an incursion. The Indians scored a public-opinion triumph when they released the recordings to international media in June 1999.[83] Two years later, after Pakistan-based terrorists attacked the Indian parliament, Indian police relied heavily on mobile-phone evidence to build a case against a teacher at a college of Delhi University. Syed Abdul Rehman Geelani spent twenty months in prison before being acquitted by the Supreme Court of India. An activist for human rights in Kashmir, Geelani and his mobile phone were picked up by police the day after the attack. Mobile phones found on the dead attackers provided a third-person link to Geelani. 'Relying heavily on mobile phone numbers', a journalist wrote, the police 'have drawn up a case of conspiracy'.[84] Geelani's fateful mobile phone foreshadowed the experience of Muhamed Haneef, an Indian doctor wrongly arrested by Australian police following the Glasgow terrorist incident in July 2007—again, because of mobile-phone connections.[85]

Mobile phones played a large part in the most coordinated terrorist attack experienced in India, the assault on central Mumbai by terrorists trained in Pakistan who landed by sea on the night of 26 November 2008. They held out for more than three days, killed more than 160 people and wounded more than 300. They used mobile phones, both their own, and those of their victims, to organise their actions, receive instructions and send pictures back to their handlers in Pakistan to keep them informed of the carnage. Recordings made by security services became widely available on the internet after Indian investigators made items recovered from the killers, including a satellite phone, available to selected diplomats and media outlets. The recordings demonstrated how handlers far away were able to guide the young killers.[86] One transcript captured the cold-bloodedness that the distant manipulator was able to transmit over the phone.[87] 'Everything is being recorded by the media. Inflict the maximum damage. Keep fighting. Don't be taken alive', a caller said to a gunman in the Oberoi Hotel in the early hours of the three-day rampage. 'Throw one or two grenades at the Navy and police teams, which are outside', came an instruction to the gunmen inside the Taj

FOR 'WRONGDOING': 'WAYWARDNESS' TO TERROR

Mahal hotel. 'Keep two magazines and three grenades aside and expend the rest of your ammunition', went another set of instructions to the attackers inside Nariman House, which housed an Orthodox Jewish centre.[88]

Victims used mobile phones to call for help, or, in this case, to send an MMS of some of the action to IBN Live, a Mumbai-based television studio. Later, a telling criticism of Indian television channels was their ill-considered live coverage, watched by the killers and their bosses in Pakistan and enabling them to take counter-measures. The attack, a *New York Times* report suggested, 'may be the most well-documented terrorist attack anywhere'.[89] Cell phones were essential tools for the attackers and amulets of hope for victims. When Mumbai suffered its next terrorist bombings in July 2011, the cover of *Outlook*, a leading news magazine, showed a burned and terrified victim, staring desperately at the camera and clutching his mobile phone. (See Illus. 35).[90]

Though few people, rich or poor, gave it much thought, cell phones connected them to the state more thoroughly than ever possible in the past. Such connections could be benign. They could make it easier to pay bills, receive entitlements and book services. But the power to give and to receive information also expanded the capacity for people to harm others, commit crimes and attack the state. States saw it as a duty to protect both themselves and law-abiding citizens. Following the Mumbai attack of November 2008, the Indian government amended its Unlawful Activities Prevention Act (UAPA) to make evidence collected through intercepted communications admissible in court, even though 'the experience with the Parliament attack' of 2001 had shown that 'electronic evidence was susceptible to tampering'.[91]

But to protect whom from what? The state from violence and sedition? Long-held 'family values' from the autonomy that young women and men might acquire from a cell phone? The innocent and the respectable from explicit sex and sexual violence? Vulnerable citizens from bullies and stalkers? Ordinary citizens from fraudsters? Unsuspecting young girls from scheming old men?[92] Society from big-time criminals and terrorists? Culpable states and their servants from public exposure? Most countries shared similar challenges. By the late 1990s, 80 per cent of drug dealers in the US were said to be 'using cloned mobile phones'.[93] Some countries waged war on pornography and targeted the corporations that enabled sexually explicit images to spread. BlackBerry yielded to the demands of the Indonesian government to filter out such material circulating through its mobile-phone services in 2011.[94] China's monitoring, eavesdropping and censoring of cell-phone and other digital traffic were notorious.[95] For authoritarian governments, closing down of mobile-phone systems at times of crisis was a standby, though one that could backfire.[96]

India experimented with a host of initiatives to establish mobile phone laws and cyber-security frameworks; but provisions were scattered across 'various statutes and governmental guidelines and rules'.[97] In 2012, proposals were made to establish a 'telecommunications security testing laboratory' to certify that all telecom equipment conformed to government regulations and did not harbour illegal tapping or disruptive devices. Such an organisation, however, was many months or years away from functioning. State police forces established modest mobile cyber-crime labs that attended crime scenes and collected evidence effectively.[98] Indian governments, however, faced a problem that wealthy states such as those in Japan, western Europe and North America had not solved: how to mitigate the evils that mobile phones could generate, while preserving their capacity to improve even a poor citizen's ability to take advantage of the rights of democratic citizenship.

CONCLUSION

'IT'S THE AUTONOMY, STUPID'

Mobile communication is not about mobility but about autonomy.
                                                            Manuel Castells[1]

We began this book with the fantasy of a film's opening sequence—the feet, the footwear and the phones in the rowboat in Banaras and the penthouse in Mumbai. How would our film have proceeded? The body of the book provides clues. Our film would hinge on the relationships and networks that grow from mobile phones. Imagine our elegant businessman in Mumbai to be a supercilious and superior character who chuckles at the presumption of the boatman and says off the cuff before ending the call, 'Come to Mumbai. Give me a call. We'll talk about your problems'. Our boatman, being a simple, literal man, takes this as an invitation and sets off with his family on the 1,300-kilometre (880 miles) journey to Mumbai. At once, we have a 'road movie' with family themes, no end of adventures and every scene pegged around a mobile phone and its connections. Imaginative readers, having understood the facets of life that cheap mobile telephony affects, will have no trouble filling in detail from their favourite Bollywood movies. A gun battle on Chowpatty Beach, a dawn cricket match and a wedding video on the screen of a 4G mobile might form the closing moments.

This book has tried to understand the profound change that cheap mobiles brought to life in India. Why only India? Because

India contained a complexity rivalled only by the world itself: a population of 1.2 billion, a resilient system of hierarchy founded on 'caste' and the only country that published newspapers and wrote SMSes in eleven different scripts. The intricacy of India sometimes suggested the world in miniature. We make a case therefore for a book about India on two grounds: that India is so large that it is worth knowing about for size alone; and that India is so complex that studying it helps to understand the nuances of an unjust, unequal, intensively mediated yet deeply divided world.

The mobile phone provided access to global flows of knowledge; altered local cultural practices; mobilised political and social movements; and challenged gender dynamics. These changes happened within a broader context, sometimes labelled Indian neoliberalism, characterised by an accelerating economy, ballooning consumerism and a middle class yearning for full-scale participation in the global economy. This globalising India, to which vast numbers of people were increasingly connected by their cell phones, often collided with the realities of everyday life; power structures and social practices were challenged. The mobile phone could be the cause of such contests, as well as a weapon and a prize. This is particularly evident in concerns about sexuality, gender relations and what some view as the corrupting influences of 'the West'. The cell phone embodied aspirations and anxieties of an age of rapid, fundamental change.

The proliferation of mobile phones after 2004 makes us forget the faltering arrival of the technology and the ten lean years from 1994. Indian governments and capitalists had to discover ways of accommodating political interests, bureaucratic rivalries and bottom-line essentials. To most Indians in those years, mobile phones were exotic, expensive and incomprehensible. The Great Indian Phone had to be cheap and simple, and then its cheapness and simplicity had to be explained and made available from Kanyakumari to Kashmir.

Thereafter, the effects of the cheap cell phone cascaded downwards, slowly at first and then with stunning rapidity. From January 2000 when the cost of a mobile-phone call fell from Rs 16 a minute to Rs 4, the device came within the means of ever larger numbers of people. For those companies that had fought to control Radio Frequency spectrum, the key lay in getting very small profits from a very large number of users. Thus followed the need

CONCLUSION: 'IT'S THE AUTONOMY, STUPID'

to make phones available at affordable prices, educate people about their simplicity and affordability and ensure that cell-phone services lived up to what advertisements promised. These processes required the involvement of millions of people as missionaries and mechanics. By 2010, as we have seen, Vodafone alone had more than a million sales points.

This book has tried to outline the processes and the activity that the mobile phone generated. But where does all this lead? What happens in a mass-based, mobile-phone future? A number of areas merit scrutiny. They include health, recycling and the environment, social networks, language and media, and politics and governance. Each prompts questions that will occupy scholars, policy makers and business analysts for many years.

Health

Cell phones affect 'health' in two ways. In the happy scenario that characterised our film script, they allow people to get medical aid quickly. Recall the 1960s advertisement in Chapter 1 for the virtues of Indian-made radio telephones—the soldier saved because his colleagues were able to radio-telephone for help. Mobile phones enable women in labour to get to medical centres or midwives to get to them. Transport reaches the ill and the injured. Beyond that, numerous experiments seek to develop digital features that will enable phones to do eye checks, test blood pressure and send photographs to specialists who can evaluate the seriousness of an injury or condition. There are plenty of positive stories. By 2009, for example, Microsoft had more than a dozen project reports for simple cell-phone applications to enable medical people to diagnose distant patients. These programs helped with the diagnosis of pneumonia and of disorders in pregnancy. They enabled the monitoring of diabetes, and technicians aimed to perfect a diagnostic microscope based on a cell phone that would allow a health worker to transmit sophisticated data to specialists in a distant centre.[2] 'Health applications', the World Bank declared, 'have the potential to transform health care systems in low-income economies'.[3]

Such devices were not panaceas. Skilled and well-equipped medical workers were needed to help people recover from disease and injury. But the capacity of cell phones to make such interventions more common did more than merely bring better diagnosis. As such interventions became more common in far-flung places, people might be expected to demand such attention as a right. The ubiquity and connectedness of mobile phones improved people's knowledge of services and planted a sense that it was their right to demand what they knew others had received. Change in expectations was almost as important as change in actual treatment.

The other side of the coin, however, concerns long-term health hazards from the air pollution and radiation emitted by cell phones and the 400,000 towers that enable Indian networks to operate. One school of thought suggests that it will take another generation before the effects of widespread radiation from individual handsets and cell-phone towers become known. At their bleakest, such interpretations contend that the parallels are with commodities like asbestos or thalidomide: ill effects will only be tangible when it is too late for many.

In 2006, the World Health Organization concluded that 'from all evidence accumulated so far, no adverse short or long-term health effects have been shown to occur from the R[adio] F[requency] signals produced by base stations' and that local-area networks in homes and offices produced even lower RF than base stations.[4] Others, however, were not so sure. An experiment in Punjab that placed a cell-phone handset in a colony of bees for 15 minutes a day for three months had the effect of 'wreaking havoc on the homing instinct' of the bees and leading to the disintegration of the colony. The research, published in 2010, got attention in New York, London and round the world.[5] A US law firm specialising in actions against corporations asserted 'that there is research from around the world that indicates a strong correlation between extensive cell phone use and brain tumours'[6]

The news magazine *Tehelka* focused on the possible dangers of RF from cell phones and the towers that transmitted their signals. Suggesting that Delhi's 6,000 towers meant that 'four-fifths of Delhi lives in unsafe zones', *Tehelka* claimed to have initiated 'an extensive survey of 100 spots ... in [the] public interest'.[7] It asserted that 40 of the 100 locations showed an 'extreme anomaly' and emitted radiation 'close to seven times the safe limit'. It told readers that 'EMR [electromagnetic radiation] is like slow poison' and 'in time' people 'may develop a high risk of cancer'.[8] While the gist of the campaign—that there was little checking on emissions from handsets or towers—was justified, the report's credibility was partially undermined by excited assertions and the fact that its survey partner had an interest in selling radiation-protection equipment.[9]

Companies selling radiation-protection gear sensed unease among many Indians about the radioactive effects of the towers and phones and the contradictory information about their impact. (See Illus. 36). And there were more tangible and immediate concerns, especially in urban areas. In Banaras many towers were located in people's backyards where they emitted noise and air pollution throughout the day. Electric power from the mains failed frequently, and fuming diesel engines ran the electronic equipment and the air-conditioning units most of the time. The huge increase in air pollution and the overall carbon footprint resulting from the diesel-powered mobile phone towers became a major concern. NGOs, some government departments and various industry bodies called for stricter regulations, better enforcement and the phasing out of diesel-generated power in favour of renewable energy, such as wind and solar.[10]

Mobile waste

Another sombre side of the mobile phone industry, which promised to grow darker in the future, was electronic waste or e-waste. There was little sustained research on the 'after life' of mobile phones, even though they were overtaking personal computers as the device producing the most e-waste. Cell phones were prized for metals, especially the rare metals and minerals (rare-earths) they contained (e.g., coltan); but not a lot was known about which metals were recovered, how this was done and what became of the salvaged material.[11] The various stages of collecting, sorting and extracting the valuable components in a mobile phone presented dangers to people's safety and health. Recycling processes often used crude methods that released toxic substances such as lead and cadmium into the air, soil and water—and into the bodies of the recyclers themselves.

The appetite for consumer goods, especially electronic devices, meant India produced staggering amounts of e-waste. Reports showed that India had also become a dumping ground. According to estimates by Toxic Link, a Delhi-based NGO, India was 'estimated to be generating approximately 400,000 tonnes of waste annually (computers, mobile phone and television only) and is expected to grow at a much higher rate of 10–15 per cent'.[12] A United Nations Environmental Program (UNEP) report predicted that e-waste from discarded mobile phones in India alone, would grow by 18 times its 2007 levels within a decade.[13]

Most recycling was done in informal industries located at the margins of society on the outskirts of metropolises and towns. NGOs reported that places such as the settlement of Seelampur on the northeast outskirts of Delhi were becoming centres where e-waste was collected, segregated and heated to extract precious metals—gold, silver, copper and others. Many working in the informal e-waste market belonged to disadvantaged Muslim communities. Conditions were bleak:

> The trucks are coming in honking and rattling with old computers, kids are playing gali [alley] cricket with computer monitors as wickets, women are boiling pots full of computer parts, children sitting on piles of keyboards are watching a Bollywood film. The streets are filled with entangled wires, destroyed computers, keyboards and cell phones. A scene from a futuristic dystopia? This is the hidden place where people wake up every morning to sort through the electronic trash of the world.[14]

Technical work did not enjoy the dignity found in the clinical environment of places such as Nokia Care or even among streetside repair-walas. Those who worked in the e-waste recycling industry were themselves considered the 'refuse of society', engaged in a stigmatised activity, and for Hindus, compounded by views about pollution and waste, caste relations and practices of exchange shaped by a Hindu cosmological order.[15] This side of consumer society was based on existing ideologies and recycling practices, which emphasised recovery and reuse. The *kabaadi-wala*—the junk dealer who came to your door—was a long-standing feature of Indian life. But the places where e-waste was collected and dealt with were sites of disorder, destitution and social fragmentation.[16] (See Illus. 37 and 38).

Electronic waste will proliferate as more and more Indians use and discard mobile phones and other gadgetry. Rapid technological advance in the production of mobile phones and electronic goods means that the composition of e-waste will change constantly. The politics of waste leads researchers to ponder the roles

CONCLUSION: 'IT'S THE AUTONOMY, STUPID'

that waste materials, wasting practices and 'waste actors' play in maintaining social order. Understanding the processes, dangers and lessons of this waste-work—collection, dismantling, processing, selling—merits another book.

Social networks

Of the puzzles that the mobile phone sets, perhaps the most nagging is the question of what it does to the way people interact—how they conduct their exchanges with others and organise their lives. To what extent do expectations and practices fundamentally change?

The earliest scholarly studies were struck by the dilemma that mobile phones posed for the industrial societies in which they first spread. 'I am', Rich Ling mused as he placed an order for a book from the Amazon company,

> a data point in a dispersed and unconnected aggregate of individuals who have the same profile according to Amazon. I will never meet these others. Even if I were to meet them, we would not likely hit it off socially ...[17]

Yet for some purposes he and they were now an entity. Ling struggled to explain the paradox. A business—Amazon—had been able to club him for commercial purposes with scores of people from all over the world whom he would never meet. He had been bundled into a mass, yet at the same time, the company treated him as an individual: referred to him by name, assembled his purchase order precisely and sent it to his address. All this happened without his having to be physically part of a group.[18] 'Networked individualism' was a term put into use by Manuel Castells and Barry Wellman at the beginning of the twenty-first century to try to capture the experiences that mobile telecommunications brought into the lives of increasing numbers of people. Castells and his associates wrote of 'an extraordinary strengthening of the culture of individualism' with the result that 'individualism rather than mobility is the defining social trend of the mobile society'.[19] A mobile phone owed its power and popularity to the autonomy it gave its owner—the chance to be alone with one other person—yet at the same time, to be in touch with people one might never see and across distances that previously would have been impossible to traverse.

Barry Wellman tried to conceptualise what was happening to societies as individual devices for communicating—e.g., mobile phones—came into the hands of larger and larger numbers of people. Most people, he suggested, once lived in 'little boxes'—face-to-face communities. Over time, as communications developed, many of them came to live in an environment shaped by the forces of 'glocalization'. People continued to live in 'little boxes' but added to their boxes distant connections, maintained by railways, steamships, telegraphs, telephones and aeroplanes. With the coming of global electronic communication, increasing numbers came to live in an environment of 'networked individualism', in which they were connected to hosts of people all over the world and at the same time could ignore their physical next-door neighbours as never before.[20]

Did this make people's lives different? Of course it did. But like the unhappy family in *Anna Karenina*, countries and cultures experienced difference in their own ways. In India, a lax democratic state imposed only sporadic control over basic mobile telephony. Most ordinary people could use the phone without great fear of being monitored and punished by the state. The state of course did not like this, and security officials and government departments sought ways to monitor activity deemed to be illegal and subversive.[21]

Violation of social norms, on the other hand, sometimes brought more rapid and targeted reprisals. India had many customs that the user of a mobile phone could become entangled with—rules and practices that were often more rigid than elsewhere. Relations between men and women, for example, were more inhibited than in much of Europe, Africa, China or North America. Everywhere in the world, mobile phones had the potential to disrupt such relations, but in India, it generated unprecedented (and often unanticipated) challenges to values, norms and practices. The vase that Cheeka, the Vodafone dog, threatened to knock over was bigger, more elaborate and perhaps more brittle than elsewhere.

People with cell phones acquired what often seemed a new individuality—some would call it 'atomisation'. 'It is not an individualism ... forced by the actors', Ling wrote, 'but, rather, an individualism that arises out of the direction of the social order'. People were individuals, but they were individuals in a powerful

CONCLUSION: 'IT'S THE AUTONOMY, STUPID'

current; they could conceivably swim against it, but it was easier not to try. Grewal wrote a book about such 'direction of the social order'. He called it *Network Power: the Social Dynamics of Globalization* and argued that 'the accumulation of individual decisions', led people, in a digital, globalised world, to adopt standards that became almost irresistible. In effect, millions of 'free' decisions by millions of individuals spun webs that entangled people in practices that proved hard to escape.[22] In India, such convergence on common standards hastened processes that had been going on at much slower paces—processes by which diverse local practices were slowly absorbed into more homogenised national customs. A country that worked in eleven different scripts, and recognised twenty-eight official languages, would only move in such directions idiosyncratically.

Homogenization into some sort of 'national' standard occurred at the same time that distinctive *local* practices were becoming more solid and celebrated. In a splendid essay entitled 'Music Mania in Small-town Bihar', Ratnakar Tripathy met 'a 54-year-old Bhojpuri poet' who told him that 'the technological churning has ... thrown up [artistic] forms he had last heard in childhood and long suspected to be extinct. For once', Tripathy concluded, 'you cannot blame technology for cultural extinction!'[23] How was this solidification and celebration of local forms achieved? 'In brief', Tripathy wrote, 'phones are the CD players of the day'.[24] Local musicians made and disseminated music, usually as cheap CDs, which were loaded onto mobile phones, sent to friends and Bluetoothed round the bazaars.[25] 'Suddenly', Tripathy wrote, 'music [in Bihar] was a respectable profession, and young girls could discuss their musical careers[,] something they could only dream of earlier'.[26] Cheap, readily available mobility placed a remarkable capacity for communications within reach of people to whom it would have been denied in the past because of their status, gender or poverty.

Electronic games and the practice of 'gaming' also arrived on mobile phones at a cost poor people could afford. To some, the global phenomenon of gaming was a global menace,[27] but its popularity was certain to grow in a country where *'timepass'* was a way of life for many underemployed young men.[28] In a slum in Hyderabad, fifteen out of the twenty young people interviewed in a survey had 'their first experience of the internet ... on a mobile phone', which provided 'a pathway to games, music and video'.[29] Games became a passion for many. 'I simply am mad about the games they have on these phones', a 17-year-old wage labourer told the researchers and explained how he barely paused to bathe after work before he borrowed a phone.[30] Another avid gamer said that gaming gave him 'nimble fingers' and improved his skills. The 'techie'—the technical wizard—in this group was 'an 8th grade drop-out' whose 'lack of literacy ... drove him to experiment and discover hardware and audio-visual content'.[31] The gaming and entertainment that mobile phones made possible had the effect of 'binding people and creating an informal technology hub'. This, the researchers concluded, is 'hardly developmental in any conventional sense', but 'what begins as entertainment can lead to more serious activities'.[32] Globally, electronic gaming was said to be worth US $90 billion in 2012, and the best international gamers, mostly in their teens, earned more than US $200,000 a year as part of sponsored teams that toured 'the games circuit' like tennis players or Formula 1 car drivers.[33] In the uncertain future of 4G in India, interactive gaming held out the promise of revenue for telecom companies, if only the costs could be kept low enough to let 17-year-old wage labourers participate and high enough to let companies recoup big investments.

The cheap mobile phone was not a magic wand of liberation. It was more like 'the old equalizer', the Colt .45 revolver in the Wild West: something to be struggled over and once possessed, not readily surrendered. More than 830 million people lived in rural India in 2011 and more than half of rural households owned a phone.[34] Though a potential tool for power-holders, the cell phone gave the less powerful—the 'information have-less', as one writer describes them—vast new vistas of entertainment and a chance, however slight, to even up life's odds a little.[35]

Language and media

A surprised scholar of linguistics told a journalist in 2012: 'We are getting languages where the *first writing* is not the translation of the Bible—as it has often happened—but text messages'.[36] He explained how people who grew up speaking small, unwritten languages, in danger of being smothered by the great state-supported languages of modern countries, found it relatively easy to

CONCLUSION: 'IT'S THE AUTONOMY, STUPID'

write SMS messages in their mother tongue. They took whatever script—Roman, Arabic, Cyrillic, etc.—they may have picked up during a brief schooling and used it as a code that enabled them to SMS in their mother tongue. The 160-character limit to an SMS message was an advantage: it not only permitted codes and contractions; it demanded them. There was no literary style to follow; one created one's own.

Such practices of code and contraction quickly appeared in highly literate European cultures when the mobile phone became common in the 1990s. SMS-ing in Finland caught on among adolescents from about 1997, and their message styles bore 'more resemblance to code than standard language'.[37] In Japan, *keitai* (cell phone) novels were written via text messaging, read widely across the globe and opened up new literary possibilities.[38] The language of text messages, often decried as the death of good grammar, also held the possibility of being the lifeblood of endangered languages.[39]

Because India operated in so many scripts,[40] SMS-ing proved less popular than in the Philippines, Indonesia or Europe, where the Roman alphabet was the sole standard. In India, many literate people were familiar only with the script of their mother-tongue, learned at primary school, and not with the Roman alphabet. The boatmen Doron knew in Banaras did not take to SMS; but they were comfortable with numbers and key sequences, very good at downloading songs and video clips and practised in using Bluetooth to transfer files. Cell phone manufacturers and service providers recognised and adapted to such markets. By 2012, Nokia had free fonts available for its phones in most Indian scripts, and the 'Panini Keypad', named for the great Sanskrit grammarian of antiquity, enabled people to download programs 'to write in all languages of India on the phone, fast and easily'.[41]

For speakers of tribal languages, such as Gondi, which was rarely written, there was immense attraction in being able to communicate widely through speech. Individuals and organisations searching for ways to make technology serve people's needs experimented with projects like CGNet Swara, the news-gathering and news-disseminating project in Chhattisgarh, and GramVaani, the citizens' monitoring and broadcasting project in Jharkhand. With all such projects two challenges stood out: sustainability (how to pay the bills?) and expansion (how to take an idea that worked for a few thousand people and make it work for millions?).

In 2010, India put more than 100 million daily newspapers on the streets each day, more than 45 per cent in Hindi, 3 per cent in English and the rest in nine other scripts and languages. But in the same year, India had more than 600 million mobile phone subscribers. The future of 'media organisations' lay with those who could find ways to use the cheap cell phone to convey information and also make a profit. The small-time entrepreneurs and exhibitionists who built huge followings through the mass text-messaging capacity of CupShup opened up a tempting path, but it was a path that led to 'a "state of stasis"', reported one analyst.

> There is no innovation on the technology front or in a revenue model despite millions frequenting the service. It became a 'free low-tech service' that is unable to monetise.[42]

As long as talk was cheap, nothing could equal talk's attraction for the mass of people in twenty-first century India. And to be able to connect with whom one liked, when one liked, across distance, raised possibilities unthinkable in the past.

Politics and governance

Socially, the mobile phone accelerated, widened and deepened change in India. The ability it gave to low-status people to communicate with each other and with sympathetic politicians and officials marked a profound break. In the past, there were words low-caste people should not hear and things they should not know. And to get close enough to a senior official or politician to expose wrong-doing or sloth was impossible for many people.

The mobile phone augmented bicycles, conviction and organisation to bring the Bahujan Samaj Party (BSP) to power in Uttar Pradesh (UP) in 2007 (Chapter 6). Subsequent history, however, underlined the truism that people, not technology, win elections. In the next elections in UP in 2012, the BSP, deserted by many disillusioned workers, was resoundingly defeated. The victorious party was led by a young man who liked to be photographed on a bicycle with a mobile phone clapped to one ear and who eulogised the booth-level organisation of his party. UP had more than 120,000 polling booths. The ability to connect with more than 100,000 party workers—however loosely committed a 'party worker' might be—could only happen regularly and constantly

CONCLUSION: 'IT'S THE AUTONOMY, STUPID'

through mobile phones. All parties by 2012 had mobile phones. The Indian National Congress, which devoted great effort and expense to the 2012 elections in UP, had ample technology, but equipment alone was not enough. A recipe for success required equipment to be combined with committed organisation based on human inspiration and on connections forged between leaders, workers and voters. No Indian political party in the twenty-first century could run an effective election campaign without mobile telephony. But true-believing workers had to be at each end of the phone for technology to affect voters.

The role of the cell phone in gathering 'flash mobs' and creating political protests from the overthrow of President Estrada in the Philippines in 2001 to the 'Arab springs' of 2011 was widely celebrated. But the phone was only a tool, and often a dangerous tool for those who protested and used it unguardedly. Morozov, dark in his visions, warned that naïve use of technology allowed 'others to identify the exact location of their owners … once you've used a cellphone, you are trapped'.[43] It is neither difficult nor expensive to track mobile-phone messages and conversations, as Indian politicians discovered to their embarrassment (and Iranian protesters sometimes at fatal cost). The ill-considered telephone conversation has been dangerous from the start: in 1896, a French play, *La Demoiselle du téléphone*, revolved around the story of a '"telephone girl in the execution of her duties overhearing her lover making an appointment with a music hall 'artiste'"'.[44] The efficacy of the cell phone in politics lay in its being a cheap tool, used for mundane organisational tasks in places that permitted fair elections and public activity. The phone gave groups with limited resources but strong convictions the capacity to connect, mobilise and broadcast. Such capacity was once reserved for the privileged. As a disruptive tool, the cellphone suited democratic India admirably.

For similar reasons, it suggested practical ways to improve governance—the way in which the state interacts with its citizens. A conversation in Bengalaru in 2010 suggested one of the possibilities: that it was feasible not merely to provide every Indian with an identity number and card but to provide every citizen with a mobile phone.[45] The conversation turned on the Aadhaar or Unique Identification Authority of India (UIAI) project, which aimed to give people an identity number validated through fingerprints and iris scans. If the state provided people with ration cards, voting cards and various subsidies, would it not make sense to provide citizens with a cheap tool to allow them more effective access to services to which they were entitled?

The 'phone for every citizen' idea would raise problems of cost, loss, theft, fraud, broadband capacity and human competence. But it was not fanciful. Indeed, it was so practical that in 2012, a desperate Congress Party, searching for popular promises for the 2014 elections, proposed a slogan of *'har haath main fon'* (a phone in every hand). The vow was to provide a cell phone and 200 monthly minutes of talk-time to six million households classified as Below the Poverty Line (BPL).[46] Even poor, rural people had digital dreams tied to telecommunications, as India's politicians recognised. The new government of Uttar Pradesh elected in 2012 promised a computer for every high school student, a proposal running to millions of laptops. And the state of Tamil Nadu put colour television sets in more than 300,000 low-income households between 2007 and 2011.[47] The scale of a phone-for-every-citizen was not impossible; what was in doubt were the side effects.

From the late 1990s, Indian governments embarked on computer-based programs intended to improve services to citizens. Andhra Pradesh and its Chief Minister gained global publicity for such initiatives when the US president, Bill Clinton, visited Hyderabad—dubbed Cyberabad—in March 2000. Success, however, varied greatly. One study, published in 2009, concluded that the effect of e-governance 'in strengthening planning systems and improving the delivery of services to citizens has so far been minimal'.[48] E-governance programs invariably required access to computers and basic ability to use them. The simplicity and ubiquity of the mobile phone offered greater potential to connect people regularly to state institutions. In Bihar, government officials—the 534 Block Development Officers or BDOs—were given ten projects in 2009 on which they were to report by text message each day. Their 4,000 daily messages, sent in a standard SMS template, were aggregated by a computer server and turned into a published report. In the past, 'BDOs took months to prepare a single report' which was rarely public. Now there were daily reports, available to interested citizens who could see for themselves whether the facts on the ground in their locality squared with what the local officer was reporting to superiors.[49]

CONCLUSION: 'IT'S THE AUTONOMY, STUPID'

Numerous small experiments were carried on around India and the world. But could small experiments become big, established and financially secure? It was one thing to run a voice-based news service or monitoring system for a few hundred or even a few thousand users and subscribers. But could such services be made to work effectively if there were millions of voices trying to be heard? Would they not dissolve into the cacophony that characterises much of the Internet? Analysts of this nexus between the communications device and political power often speculate about possibilities 'for new forms of social ties, organisations and behaviours' and of 'altering mindsets and behaviour'.[50] By 2010, government officers at virtually every level were required to publicise their mobile phone numbers.[51] Nonetheless, as a short story entitled 'Mobile Phone' in a Hindi newspaper suggested, a person was likely to increase her chances of a reply from a local official, if the official's pre-paid talk time was topped up handsomely in advance by the petitioner—a digital bribe.[52] However, the possibility that such an illegal exaction could be publicised represented a new avenue by which citizens could make the state fulfil its promises and carry out its duties. The greater potential of individuals to record, report and broadcast could enforce new standards of probity and conscientiousness. Officers facing temptation might conclude: because we might be detected, we will behave. The cheap phone provided opportunity to organise, broadcast and appeal to the law.

In some circumstances, too, 'unprecedented access to information and resources'[53] improved the earning capacity of farmers and fisherfolk, rickshaw drivers, boatmen and *dhobis*—not to say the earning capacity of those who controlled Airtel, Reliance, Tata, Vodafone, Nokia and others.[54] A host of new occupations grew up around the mobile-phone industry—recyclers, repairers, second-hand dealers, sales people, advertising agents, marketers, technicians, executives, manufacturers and builders. Some of the new occupations widened opportunities.

Fernand Braudel wrote of the 'boundary between possibility and impossibility ... between what can ... be attained and what remains denied' to people at particular times in history and of the importance of the moment when there is 'an extension of the limit of possibility'.[55] The cell phone provides such an extension. It makes its owners into potential publishers and documentary makers. It surmounts physical barriers—'leaps tall buildings at a single bound'. A person's voice can bypass the oppressive landlord or unsympathetic policeman, and an individual can arrange a meeting, tell a story, warn a friend, lodge a complaint or call for help without having either to be present or literate. Violations of the law can be photographed and disseminated without the knowledge of perpetrators that their actions have been captured.

The cell phone drew India's people into relations with the record keeping capitalist state more comprehensively than any previous mechanism or technology. By itself, none of this overturned power structures or ironed out inequality, yet, as Braudel wrote of the improvements in road transport 180 years ago, it did make conditions 'faster, more efficient and'—a matter of hope and promise—'more democratic'.[56]

NOTES

PREFACE

1. At the time, the rupee was worth about 29 to the USD, but throughout the book we use a rough calculation of Rs 50 to US $1.00, which was where the rupee hovered in 2010.

INTRODUCTION: 'SO UNCANNY AND OUT OF PLACE'

1. While the Anglicised name is the River Ganges, we use the name Ganga as the river is popularly known. Varanasi is the official name for the city, though the name Banaras is equally popular and we use both names throughout the book.
2. We use 'mobile' and 'cell' phone interchangeably as the terms commonly are in India.
3. Vishnu Bhatt and Godshe Versaikar, *1857. The Real Story of the Great Uprising*, trans. Mrinal Pande (New Delhi: Harper Perennial, 2011), p. 205.
4. Giogio Riello and Peter McNeill, 'A Long Walk: Shoes, People and Places', in Giogio Riello and Peter McNeill (eds), *Shoes: a History from Sandals to Sneakers* (New York: Berg, 2006), p. 12.
5. But even shoes were not that common. A survey in 1989 estimated that Indians acquired one new pair of shoes or sandals every two years. *Business India*, 17–30 April 1989, p. 87.
6. Allen Shaw, 'History of the Watch', http://reference.arama.com/fashionstyle/45078.php (accessed 28 June 2012).
7. Telecom Regulatory Authority of India [TRAI], 'Information Note to the Press (Press Release No. 51/2011', 20 October 2011, p. 1.
8. *Traffic and Market Report on the Pulse of the Networked Society* (Stockholm: Ericsson, 2012), p. 5, http://www.ericsson.com/res/docs/2012/traffic_and_market_report_june_2012.pdf (accessed 3 July 2012).

9. *Information and Communications for Development 2012: Maximizing Mobile* (Washington, DC: World Bank, 2012), p. xi, http://www.worldbank.org/ict/IC4D2012(accessed 31 July 2012).
10. Russell Southwood, *Less Walk, More Talk: How Celtel and the Mobile Phone Changed Africa* (Chichester: Wiley for Celtel, 2008), p. xv. This commissioned history of the company is well-written and researched.
11. Ibid., pp. 203–04.
12. *Economic Times*, 4 November 2011, http://economictimes.indiatimes.com/articleshow/10591762.cms?prtpage=1 (accessed on 4 November 2011). *International Herald Tribune*, 10 May 2011, p. 15. *Economist*, 4 February 2012, pp. 52–3. *Observer*, 1 February 2009, www.guardian.com.uk/lifestyle/2009/feb/01/mo-ibrahim/print (accessed 1 March 2012).
13. *People's Daily Online*, 26 April 2011, http://english.peopledaily.com.cn/90001/90776/90882/7361295.html (accessed 26 April 2011).
14. Tamar Ashuri, *From the Telegraph to the Computer: A History of Electronic Media* (Tel-Aviv: Resling Publishing, 2011, in Hebrew), pp. 30–31. See also David Edgerton, *The Shock of the Old: Technology and Global History since 1900* (Oxford: Oxford University Press, 2006); Carolyn Marvin, *When Old Technologies Were New: Thinking about Electronic Communication in the Late Nineteenth Century* (Oxford: Oxford University Press, 1988); Chandrika Kaul, *Reporting the Raj* (Manchester: Manchester University Press, 2004); and Deep Kanta Lahiri Choudhury, *Telegraphic Imperialism* (Houndmills: Palgrave Macmillan, 2010).
15. Sukh Ram, 1993–6 (convicted and imprisoned); Beni Prasad Verma, 1996–8; Buta Singh, 1998; Sushma Swaraj, 1998; Jagmohan, 1998–9 (removed); Ram Vilas Paswan, 1999–2001 (Minister of State); Pramod Mahajan, 2001–03 (murdered); Arun Shourie, 2003; Dayanidhi Maran, 2004–07; A. Raja, 2007–10 (charged).
16. John Powell, *The Survival of the Fitter: Lives of Some African Engineers* (Rugby: Intermediate Technology Publications, 1995), pp. 114–15.
17. Telecom Regulatory Authority of India [TRAI], *Annual Report, 2010–11*(New Delhi: TRAI, 2011), p. 4.
18. 'Wala' denotes someone engaged in the business of 'x' or in this case, repairing things.
19. Leopoldina Fortunati, Anna Maria Manganelli, Pui-lam Law and Shanhua Yang, 'Beijing Calling ... Mobile Communication in Contemporary China', *Knowledge, Technology and Policy*, vol. 21, no. 1 (2008), p. 26.
20. *Census of India 2011. Houses. Household Amenities and Assets. Latrine Facility*, http://www.censusindia.gov.in/2011census/hlo/Data%20sheet/Latrine.pdf (accessed 16 April 2012). For HP, *Times of India*, 2 April 2012, http://timesofindia.indiatimes.com/articleshow/12499306.cms (accessed 3 April 2012). A UN report in 2010 first highlighted the contrast between phone and toilet ownership. 'Mobile Phones More Common than Toilets in India,

UN Report Finds', 14 April 2010, http://www.un.org/apps/news/printnewsAr.asp?nid=34369 (accessed 16 April 2012).
21. C. K. Prahalad, *The Fortune at the Bottom of the Pyramid: Eradicating Poverty through Profits* (Upper Saddle River, NJ: Wharton School Publishing, 2005), p. 24.
22. Jeffrey is happy to present his Australian phone bills in evidence.
23. Madhu Trehan, *Tehelka as Metaphor* (New Delhi: Roli, 2009). 'Tehelka report was not entirely false: Arun Jaitley', *Times of India*, 1 February 2009, http://articles.timesofindia.indiatimes.com/2009-02-01/india/28047183_1_jayajaitley-bangaru-laxman-george-fernandes (accessed 6 December 2011).
24. The full Idea ad, called 'Fearless Life', was on the Idea website in April 2012. A shorter version was on Youtube. http://www.youtube.com/watch?v=fO-SMCnbic (accessed 19 June 2012).
25. Shirin Madon, *e-Governance for Development: a Focus on Rural India* (Houndmills: Palgrave Macmillan, 2009), p. 94,
26. Clay Shirky, *Here Comes Everybody: How Change Happens When People Come Together* (London: Penguin, 2008), pp. 143–56.
27. Manuel Castells, *The Rise of the Network Society*, 2nd edition (Chichester: Wiley-Blackwell, 2010; first published 1996), p. xviii.
28. Ibid., p. 500. We discussed some of these issues in a different context in Robin Jeffrey and Assa Doron, 'The Mobile Phone in India and Nepal: Political Economy, Politics and Society', *Pacific Affairs*, vol. 85, no. 3 (September 2012), pp. 469–82.
29. Heather Horst and Daniel Miller, *The Cell Phone: an Anthropology of Communication* (Oxford: Berg, 2006). Rich Ling, *The Mobile Connection: the Cell Phone's Impact on Society* (San Francisco: Elsevier, 2004). James E. Katz (ed.), *Handbook of Mobile Communication Studies* (Cambridge, MA: MIT Press, 2008).
30. James E. Katz and Mark Aakus (eds), *Perpetual Contact* (Cambridge: Cambridge University Press, 2002).
31. See, for example, Stephanie, H. Donald, Theresa D. Anderson and Damien Spry (eds.) *Youth, Society and Mobile Media in Asia* (London: Routledge, 2010).
32. Claude S. Fisher, *America Calling: a Social History of the Telephone* (Berkeley, CA: University of California Press, 1994), p. 85.
33. Sadie Plant, 'On the Mobile: the effects of mobile telephones on social and individual life', http://classes.dma.ucla.edu/Winter03/104/docs/splant.pdf (accessed 10 July 2012).
34. Mizuko Ito, Daisuke Okabe and Misa Matsuda (eds), *Personal, Portable, Pedestrian: Mobile Phones in Japanese Life* (Cambridge, MA: MIT Press, 2005). The 15 essays focus particularly on young people's use of mobiles in Japan. Chapters 2, 4, 5 and 14 deal explicitly with young people.
35. Inge Brinkman, Mirjam de Bruijn and Hisham Bilal, 'The Mobile Phone,

"Modernity" and Change in Khartoum, Sudan', in Mirjam de Bruijn, Francis Nyamnjob and Inge Brinkman (eds), *Mobile Phones: the New Talking Drums of Everyday Africa* (Leiden: African Studies Centre, 2009), p. 81. Heather A. Horst and Daniel Miller, *The Cell Phone: an Anthropology of Communication* (Oxford: Berg, 2006), p. 57.

36. Francis B. Nyamnjoh, 'Married But Available', excerpt in de Bruijn, Nyamnjob and Brinkman (eds), *Mobile Phones*, p. 3.
37. Daniel Miller and Don Slater, 'Comparative Ethnography of New Media', in James Curran and Michael Gurevitch (eds), *Mass Media and Society*, 4th edition (London: Hodder Arnold, 2005), p. 309.
38. Brinkman *et al.*, 'The Mobile Phone', p. 88. This approach is designed to counter 'techonological determinism': a view that technological change regulates and drives both individual activities and wider cultural and economic structures.
39. Johan Fischer, *Shopping among the Malays in Modern Malaysia* (Copenhagen: Nordic Institute of Asian Studies, 2008), pp. 29–32, 106.
40. http://www.ameinfo.com/43982.html (accessed 22 March 2012)
41. Speech-coding technology allows operators to carry more traffic by reducing the richness of the signals. This may reduce the quality of voice transmission but allows networks to handle many more SMS messages.
42. Omri Shamir and Guy Ben-Porat, 'Boycotting for Sabbath: Religious Consumerism as a Political Strategy', *Contemporary Politics*, vol. 13, no. 1 (2007), pp. 75–92.
43. Steven Erlanger, 'A Modern Marketplace for Israel's Ultra-Orthodox', *New York Times*, 2 November 2007, http://www.nytimes.com/2007/11/02/world/middleeast/02orthodox.html?_r=2&oref=slogin&oref=slogin (accessed 14 April 2012).
44. Mukesh K. Singh and Vasim Ansari, 'dharm ka granth' (the book of dharma). *My Mobile Magazine* (in Hindi), (September, 2011), pp. 15–17. See also Phyllis Herman, 'Seeing the Divine through Windows: Online Darshan and Virtual Religious Experience', *Heidelberg Journal of Religions on the Internet*, vol. 4, no. 1 (2010), pp. 151–178.
45. Dennis McGilvray, email to A. Doron, 8 June 2012.
46. Bart Barendregt, 'Mobile Religiosity in Indonesia: Mobilized Islam, Islamized Mobility and the Potential of Islamic Techno Nationalism', in Erwin Alampey (ed.), *Living the Information Society in Asia* (Singapore: Institute of Southeast Asian Studies, 2009), p. 80.
47. For successful efforts by the Indonesian government to ban access to pornography via BlackBerry services, see http://www.thestar.com/business/companies/rim/article/919345-rim-to-censor-web-porn-on-blackberrys-in-indonesia (accessed 22 March 2012).
48. Jia Lu and Ian Weber, 'State, Power and Mobile Communication: A Case Study of China', *New Media Society*, vol. 9, no. 6 (2007), p. 936.

NOTES pp. [13–22]

49. Jack Linchuan Qiu, *Working-Class Network Society: Communication Technology and the Information Have-Less in Urban China*(Cambridge, MA: MIT Press, 2009), p. 188.
50. Lu and Weber, 'State, Power and Mobile Communication', p. 937.
51. See, for example, Hiyam Omari-Hijazi and Rivka Ribak, 'Playing with Fire: On the domestication of the mobile phone among Palestinian teenage girls in Israel', *Information, Communication & Society*, vol. 11, no. 2 (2008), pp. 149–166. Eija-Liisa Kasesniemi and Pirjo Rautianinen, 'Mobile Culture of Children and Teenagers in Finland', in Katz and Aakhus (eds), *Perpetual Contact*, pp. 170–92.
52. Marvin, *When Old Technologies Were New*, p. 70.
53. Charles Thompson, Manager, Customer Relations, EKO, New Delhi, email to R. Jeffrey, 6 July 2012.

1. CONTROLLING COMMUNICATION

1. Marc Bloch, *Feudal Society*, vol. 1, *The Growth of Ties of Dependence* (London: Routledge and Kegan Paul, 1961), p. 64 and the preceding discussion and pp. 93–4.
2. Annemarie Schimmel, *The Empire of the Great Mughals* (London: Reaktion Books, 2004), p. 33.
3. Michael H. Fisher, 'The Office of Akhbar Nawis: the Transition from Mughal to British Forms', *Modern Asian Studies*, vol. 27, no. 1(February 1993), p. 50.
4. Niccolao Manucci, *Mogul India, 1653–1708*, vol. 2, trans. William Irvine (London: John Murray, 1907), pp. 331–2.
5. Fisher, 'Office of Akhbar Nawis', p. 54.
6. Schimmel, *Empire of the Great Mughals*, p. 101.
7. Vishnu Bhatt and Godshe Versaikar, *1857. The Real Story of the Great Uprising*, trans. Mrinal Pande (New Delhi: Harper Perennial, 2011; first publish 1908), p. 11.
8. R. K. Narayan, *The Grandmother's Tale* (London: Heinemann, 1993), p. 24 and thereafter. Velcheru Narayana Rao and Sanjay Subrahmanyam, 'Circulation, Piety and Innovation: Recounting Travels in Early Nineteenth-Century South India', *Society and Circulation* (New Delhi: Permanent Black, 2003), pp. 309–10, discuss a documented tale similar to the one in Narayan's novel.
9. Narayan, *Grandmother's*, pp. 42–3.
10. Ibid., p. 22.
11. Quoted in Fisher, 'Office of Akhbar Nawis', p. 74.
12. Fisher, 'Office of Akhbar Nawis', p. 57. For an excellent account of intelligence and spying throughout history, Rahul Sagar, 'The Institutionalization and Centralization of Secret Intelligence', unpublished manuscript.
13. Claude Markovits, 'Merchant Circulation in South Asia', in Claude Mar-

kovits, Jacques Pouchepadass and Sanjay Subrahmanyam (eds), *Society and Circulation* (London: Anthem Press, 2006), p. 156.
14. Jean Deloche, *Transport and Communications in India Prior to Steam Locomotion*, vol. 1, *Land Transport*, trans. James Walker (New Delhi: Oxford University Press, 1993), p. 217.
15. Barbara Daly Metcalf, *Islamic Revival in British India: Deoband, 1860–1900* (Princeton, NJ: Princeton University Press, 1982), p. 62.
16. Bloch, *Feudal Society*, vol. 1, p. 62.
17. Patrick Olivelle (ed. and trans.), *Dharmasutra. The Law Codes of Ancient India* (Oxford: Oxford University Press, 1999), 12.1, p. 98.
18. Sudha Pai, *Dalit Assertion and the Unfinished Democratic Revolution* (New Delhi: Sage, 2002), p. 37.
19. *Travancore Government Gazette*, vol. 32, no. 45 (6 November 1895), advertisement seeking a replacement priest.
20. Eric J. Miller, 'Caste and Territory in Malabar', *American Anthropologist*, vol. 56, no. 3 (June 1954), p. 419.
21. Deloche, *Transport and Communications*, vol. 1, p. 220.
22. Miller, 'Caste and Territory', p. 415. William Logan, *Malabar*, vol. 1 (Madras: Government Press, 1887; reprint 1951), p. 129.
23. *Wired*, 1 December 2008, http://www.wired.com/dangerroom/2008/12/the-gagdets-of/ (accessed 18 January 2012).
24. Quoted in Fisher, 'Office of Akhbar Nawis', p. 64.
25. Bernard S. Cohn, 'The Command of Language and the Language of Command', in Bernard S. Cohn, *Colonialism and Its Forms of Knowledge* (Princeton: Princeton University Press, 1996), p. 16.
26. B. S. Kesavan, *History of Printing and Publishing in India*, vol. 1, *South Indian Origins of Printing and Its Efflorescence in Bengal* (New Delhi: National Book Trust of India, 1985), p. 209.
27. Ernest Sackville Turner, *The Shocking History of Advertising* (New York: Ballantine Books, 1953), p. 17.
28. John Lang, *Legends of India. Tales of Life in Hindostan*, ed. by Victor Crittenden (Canberra: Mulini Press, 2008), p. 58. At the end of the nineteenth century, one pound sterling was equivalent to about 15 rupees.
29. Lang, *Legends*, pp. 62–3.
30. Deep Kanta Lahiri Choudhury, *Telegraphic Imperialism* (Houndmills: Palgrave Macmillan, 2010), pp. 139–43, has an account of the Vernacular Press Act, though it has the wrong Viceroy assassinated. It was Mayo, not Minto.
31. Lahiri Choudhury, *Telegraphic Imperialism*, pp. 21–22, 40
32. C. A. Bayly, *Empire and Information* (Cambridge: Cambridge University Press, 1996), p. 317.
33. The remark is often repeated; the original source is unclear. John H. Lienhard, 'Indian Telegraph', offers a concise account, http://www.uh.edu/engines/epi1380.htm (downloaded 14 October 2010).

34. Bhatt and Versaikar, *1857*, p. 29.
35. Horst and Miller, *The Cell Phone*, p. 105.
36. Herbert N. Casson, *The History of the Telephone* (New York: Cosimo Classics, 2006; first published 1910), p. 23.
37. *Travancore Administration Report, 1878–9* (Trivandrum: Government Press, 1879), p. 78.
38. *Forty Years of Telecommunications in Independent India* (New Delhi: Department of Telecommunications, n.d. [1987]), p. v.
39. *Forty Years*, p. 245.
40. M. K. Gandhi to Munnalal G. Shah, 20 December 1941, No. 646, *Collected Works of Mahatma Gandhi*, vol. 81, p. 388.
41. M. K. Gandhi, *Hind Swaraj* (many editions since 1915), p. 96.
42. Gandhi to Shah, 20 December 1941, in *CWMG*, vol. 81, p. 388.
43. *Forty Years*, p. 2.
44. *Times of India*, 13 November 1955. The Communications Minister was the legendary Jagjivan Ram (1908–86).
45. *Economic Weekly*, 14 May 1960, p. 727.
46. *Economic and Political Weekly*(hereafter *EPW*), 20 June 1970, p. 961.
47. *EPW*, 21 November 1987, p. 1987.
48. *Statistical Outline of India, 2000–01* (Mumbai: Tata Services, 2000), p. 78.
49. *India: a Reference Annual, 1964* (New Delhi: Publications Division, 1964), pp. 329, 307, 332.
50. *India 1964*, pp. 121, 125.
51. Nalin Mehta, *India on Television* (New Delhi: HarperCollins, 2008), pp. 27–40.
52. Peter Manuel's wonderful book, *Cassette Culture* (Chicago: University of Chicago Press, 1993) and the use made of cassettes by the *gurbani* reciting preacher-politician Jarnail Singh Bhindranwale. Robin Jeffrey, *What's Happening to India?* (London: Macmillan, 1986), p. 92.
53. *India. A Reference Annual 1953* (New Delhi: Publications Division, 1953), p. 332.
54. The trailer was at http://www.youtube.com/watch?v=WlA665YP_Ww on 19 March 2012.
55. Department of Posts, *Annual Report* (New Delhi: Department of Posts, India, for relevant years).
56. Ibid., p. Ad[vertisement] 19.
57. For a description of the caste and gender anxieties provoked by the arrival of the railways to India, see Tanika Sarkar, *Hindu Wife, Hindu Nation: Community, Religions, and Cultural Nationalism* (London: C. Hurst, 2001), pp. 81–2.
58. Olivelle (ed. and trans.), *Dharmasutra*, 12.1, p. 98.
59. Robin Jeffrey, 'The Mahatma Didn't Like the Movies and Why it Matters: Indian Broadcasting Policy, 1920s–1990', in Robin Jeffrey, *Media and Modernity* (New Delhi: Permanent Black, 2010), pp. 241–2.

60. *First Five-Year Plan*, Chapter 31, section 7, http://planningcommission.nic.in/plans/planrel/fiveyr/welcome.html (downloaded 13 October 2010).
61. *Second Five-Year Plan*, Chapter 22, para 12, http://planningcommission.nic.in/plans/planrel/fiveyr/welcome.html (downloaded 13 October 2010).
62. Ashok V. Desai, *India's Telecommunications Industry* (New Delhi: Sage, 2006), p. 41.
63. Heather Horst and Daniel Miller, 'From Kinship to Link-up: Cell Phones and Social Networking in Jamaica', *Current Anthropology*, vol. 46, no. 5 (December 2005), p. 762.
64. Mira Kamdar, *Planet India* (New York: Scribner, 2007), p. 106.
65. Quoted in Rafiq Dossani, *India Arriving* (New York: American Management Association, 2008), p. 49.
66. Varadharajan Sridhar, *The Telecom Revolution in India: Technology, Regulation and Policy* (New Delhi: Oxford University Press, 2012), p. 22.
67. Population estimated on 2 per cent annual increase from the 1981 census population of 683 million people. Phone numbers from *Forty Years*, p. 89.
68. T. Hanuman Chowdhary, 'Rajiv Gandhi—Promoter of I.T. and Public Sector Efficiency', http://www.drthchowdhary.net (downloaded 22 October 2010). Chowdhary was the first managing director of VSNL.
69. S. D. Saxena, *Connecting India: Indian Telecom Story* (New Delhi: Konark, 2009), p. 46.
70. Dilip Subramanian, *Telecommunications Industry in India* (New Delhi: Social Science Press, 2010), p. 41.
71. Desai, *India's Telecommunications Industry*, p. 43.
72. Saxena, *Connecting India*, p. 58.
73. Sam Pitroda, 'Development, Democracy and the Village Telephone', *Harvard Business Review*, vol. 71, no. 6, November-December 1993, pp. 66–79. http://groups.google.com/group/soc.culture.indian/browse_thread/thread/c0e823d72686cd64/6bb8fa00c1d2790f (downloaded 22 October 2010).
74. Paula Chakravartty, 'Telecom, National Development and the Indian State: a Postcolonial Critique', *Media, Culture and Society*, vol. 26, no. 2 (2004), pp. 241–3. Saxena, *Connecting India*, p. 66.
75. Subramanian, *Telecommunications Industry in India*, pp. 113–15.
76. Pitroda, 'Development', (downloaded 22 October 2010).
77. Desai, *India's Telecommunications Industry*, p. 42.
78. Agar, *Constant Touch*, pp. 29–102.
79. *EPW*, 20 June 1970, p. 961.
80. 'National Telecom Policy 1994', http://www.trai.gov.in/Content/telecom_policy_1994.aspx (accessed 6 August 2012). 'The defence and security interests of the country will be protected'.
81. Ibid.
82. Ibid.

83. Agar, *Constant Touch*, p. 40.
84. Manucci, *Mogul India*, p. 421.
85. Vinod Mehta, *Lucknow Boy* (New Delhi: Penguin/Viking, 2011), p. 259.
86. The Duke of Cumberland led the English armies that suppressed the Scottish Highlands after the failed revolt of 1745–6.

2. CELLING INDIA

1. Barbara Crossette in *New York Times*, 22 May 1991, http://articles.orlandosentinel.com/1991-05-22/news/9105220915_1_gandhi-rajiv-chandrashekhar (accessed 16 December 2010).
2. Christopher Kremmer, email to R. Jeffrey, 16 December 2010. Kremmer was the Australian Broadcasting Corporation correspondent in India in 1991.
3. *Forty Years of Telecommunications in Independent India* (New Delhi: Department of Telecommunications, n.d. [1988]), p. 40.
4. Desai, *India's Telecommunications Industry*, p. 19.
5. *TRAI Annual Report, 2008–09*, p. 40 and subsequent *TRAI ARs*.
6. Muthuswamy and Brinda (eds) [no initials], *Swamy's Treatise on Telephone Rules with Act* [sic], *Rules, Orders, Digest, Guidelines and Case-Law* (Madras: Swamy Publishers, 1993; first published 1989), pp. 10–11. The new era of mobile-phone private enterprise generated a 730-pager—Pavan Duggal, *Mobile Law* (New Delhi: Universal Law Publishing Co. Pvt Ltd, 2011).
7. N. Sreekantan Nair, Thiruvananthapuram, interview with R. Jeffrey, 20 November 2010. Mr Sreekantan Nair was a senior officer in telecommunications for the Government of India from 1966 to 2004. Daniel E. Sullivan, John L. Sznopek and Lorie A. Wagner, *20th Century U.S. Mineral Prices Decline on Constant Dollars* (n.p.p.: U.S. Geological Survey, Open File Report 00–389, n.d.), p. 5, http://www.fxstreet.com/fundamental/economic-time-series/data/fedstl/exinus.aspx (accessed on 19 March 2012). Copper was worth about US $8 a kilo in New York in 2012. Bloomberg, 16 March 2012, http://www.bloomberg.com/news/2012-03-16/copper-pares-weekly-gain-on-rising-china-stockpiles-correct-.html (accessed 18 June 2012).
8. Desai, *India's Telecommunications Industry*, p. 42.
9. M. B. Athreya, 'India's Telecommunications Policy: a Paradigm Shift', *Telecommunications Policy*, vol. 20, no. 1 (1996), pp. 16–17.
10. This eventually happened with the creation of Bharat Sanchar Nigam Ltd (BSNL) in 2000.
11. Athreya, 'India's', p. 17.
12. Ibid., p. 18.
13. Desai, *India's Telecommunications Industry*, p. 47.
14. Deepali Sharma and Abhoy K. Ojha, 'Evolution of the Indian Mobile Telecommunications Industry: Looking through the C-Evolutionary Lens', paper given at the 'Celling South Asia' workshop, Institute of South Asian

Studies, Singapore, 17 and 18 February 2011. Mukesh Kumar and Ram Kumar Kakani, *The Telecommunications Revolution: Mobile Value Added Services in India* (New Delhi: Social Science Press, 2012), pp. 12–19.
15. Sharma and Ojha, 'Evolution', workshop paper. Arun K. Thiruvengadam and Piyush Joshi, 'Judiciaries as Crucial Actors in Southern Regulatory Systems: A Case Study of Indian Telecom Regulation', *Regulation and Governance*, 2012, pp. 8–9.
16. TRAI, *National Telecom Policy 1994*.
17. These roughly corresponded with the states of India's federation, but small states—in the northeast, for example—were grouped into single circles.
18. Athreya, 'India's', p. 13.
19. William H. Melody, 'Spectrum Management for Information Societies', in John Ure (ed.), *Telecommunications Development in Asia* (Hong Kong: Hong Kong University Press, 2008), p. 72.
20. Peter D. O'Neill, 'The "Poor Man's Mobile Telephone": Access versus Possession to Control the Information Gap in India', *Contemporary South Asia*, vol. 12, no. 1 (2003), p. 89.
21. Prabhir Purkayastha, 'Induction of Private Sector in Basic Telecom Services', *EPW*, 17 February 1996, p. 417.
22. Purkayastha, 'Induction', p. 419.
23. Desai, *India's Telecommunications Industry*, p. 47.
24. Ibid., p. 48.
25. Ibid., p. 49.
26. Praveen R. Kumar, Nokia Siemens Network, Thiruvananthapuram, interview with R. Jeffrey, 21 November 2010.
27. Anupama Dokeniya, 'Re-forming the State: Telecom Liberalization in India', *Telecommunications Policy*, vol. 23 (1999), p. 115.
28. Dokeniya, 'Re-forming', p. 122.
29. *Economic Times*, 7 April 2001, www.valuenotes.com/et01/apr07.asp?ArtCd=24724&Cat=C&Id=69 (accessed 29 March 2009). It sold its stake to the Hindujas.
30. *Indian Express*, 21 February 2009, www.indianexpress.com/news/assets-case-takes-13-years-for-sukh-rams-conviction (accessed 29 March 2009).
31. He was sent to hospital in June with 'multiple ailments'. Indo-Asian News Service, 4 June 2012, http://gujaratinews.webdunia.com/english-news/shownews/0/Sukh-Ram-shifted-to-hospital/12867314.html (accessed 20 June 2012).
32. Desai, *India's Telecommunications Industry*, p. 55.
33. *Indian Express*, 15 December 2007, www.expressindia.com/story_print.php?storyId=250527 (accessed 19 February 2010).
34. Desai, *India's Telecommunications Industry*, p. 55. Rahul Mukherji, 'Interests, Wireless Technology and Institutional Change: From Government Monopoly to Regulated Competition in Indian Telecommunications', *Journal of Asian Studies*, vol. 66, no. 2 (May 2009), p. 498.

35. Mukherji, 'Interests', pp. 501–02. Thiruvengadam and Joshi, 'Judiciaries', p. 9.
36. Desai, *India's Telecommunications Industry*, p. 52.
37. S. S. Sodhi in *Financial Express*, 23 August 1998, www.financialexpress.com/old/fe/daily/19990823/fec23033p.html (accessed 3 March 2011).
38. The Communications Ministers were Sukh Ram, Beni Prasad Verma, Buta Singh, Sushma Swaraj, Jagmohan and Ram Vilas Paswan. Prime Minister Atal Bihari Vajpayee took over the portfolio for a few months in 1999.
39. *Statistics of India* and *India: a Reference Annual* for relevant years.
40. Desai, *India's Telecommunications Industry*, pp. 114–15.
41. Ibid., p. 84.
42. *Rediff on the Net*, 9 June 1999, quoting Jagmohan, the former minister. Mahesh Uppal with S. K. N. Nair and C. S. Rao, *India's Telecom Reform: a Chronological Account* (New Delhi: National Council for Applied Economic Research, 2006), p. 5.
43. Desai, *India's Telecommunications Industry*, pp. 79–80.
44. *New Telecom Policy 1999 (NTP 1999)*, http://www.trai.gov.in/TelecomPolicy_ntp99.asp (accessed 4 March 2011). The 1994 policy was officially called 'national'; the 1999 policy was officially dubbed 'new'.
45. Rafiq Dossani (ed.), *Telecommunications Reform in India* (London: Quorum Books, 2002), p. 5. Kumar and Kakani, *The Communications Revolution*, pp. 15–18.
46. Desai, *India's Telecommunications Industry*, p. 141. Mukherji, 'Interests', p. 505.
47. Desai, *India's Telecommunications Industry*, p. 88. Sridhar, *Telecom Revolution*, p. 82.
48. Saxena, *Connecting India*, pp. 3–47.
49. Uppal et al., *India's Telecom Reform*, p. 6.
50. Mukherji, 'Interests', p. 507.
51. Desai, *India's Telecommunications Industry*, p. 30. Sridhar, *Telecom Revolution*, p. 19.
52. The numbers are no doubt inflated by up to 30 per cent. This is because a person who lets an old number lapse and buys a new number is counted twice. Nevertheless, the increase of cell-phone penetration, even allowing for such inflation, was immense.
53. Hamish McDonald, *Mahabharata in Polyester* (Sydney: University of New South Wales Press, 2010), p. 303.
54. This was called CDMA (Code Division Multiple Access) and involved different technology from that adopted by much of the rest of the world and by the mobile-service providers licensed in India in 1995—GSM (Global System for Mobile communications).
55. Arun Shourie, *Governance and the Sclerosis That Has Set In* (New Delhi: Rupa, 2007; first published 2004), pp. 77–8. McDonald, *Mahabharata*, pp. 304–05. Desai, *India's Telecommunications Industry*, p. 100.

56. McDonald, *Mahabharata*, pp. 303–09 and Uppal et al., *India's Telecom Reform*, pp. 8–10 detail this complicated story. Thiruvengadam and Piyush Joshi, 'Judiciaries', pp. 11–12.
57. Agar, *Constant*, pp. 39–40.
58. Ibid., p. 62.
59. Melody, 'Spectrum', pp. 78–9.
60. *TRAI Annual Report, 2009–10*, p. 28.
61. Pankaj Misra, *Butter Chicken in Ludhiana* (London: Picador, 2006; first published 1995). Anand Giridharadas, *India Calling* (Melbourne: Black Ink., 2011), pp. 211 ff.
62. *Outlook*, 16 October 2006, www.outlookindia.com/printarticle/aspx?232842 Z (accessed 18 September 2009).
63. Sunil Bharti Mittal, *India's New Entrepreneurial Classes: the High Growth and Why It Is Sustainable*, Occasional Paper No. 25 (Philadelphia: Centre for the Advanced Study of India, University of Pennsylvania, 2006), p. 15.
64. F. Asis Martinez-Jerez and V. G. Narayanan, 'Strategic Outsourcing at Bharti Airtel Limited', *Harvard Business School 9–107–003*, 2006 (revised 4 December 2007), p. 2.
65. *Straits Times* (Singapore), 22 March 2010, p. B21. *International Herald Tribune*, 13 January 2010, p. 14.
66. Desai, *India's Telecommunications Industry*, p. 25. Mukherji, 'Interests', p. 509.
67. Quoted in Keskar, 'Reliance Infocomm', p. 199.
68. Ibid., p. 206.
69. *TRAI Annual Report, 2009–10*, pp. 28–9.
70. www.forbes.com/lists/2006/10/OVAG.html (accessed 8 March 2011).
71. *TRAI Annual Report, 2009–10*, p. 28.
72. In 2002 Videsh Sanchar Nigam Ltd (VSNL), created as a government corporation to take over telecommunications in 1986, was privatised. Tata became the main shareholder. C. N. N. Nair, *The Story of Videsh Sanchar: Development of India's External Telecommunications* (Mumbai: Videsh Sanchar Nigam Ltd, 2002), p. 131.
73. *TRAI Annual Report, 2009–10*, p. 28. BSNL was estimated at about 12 per cent and Idea and Tata at about 11 per cent each.
74. www.adityabirla.com/our_companies/indian_companies/idea.htm (accessed 10 March 2011).
75. Suneeta Reddy in *Financial Express*, 11 March 2006, www.financialexpress.com/printer/news/42185 (accessed 10 March 2011). The Aircel deal was the subject of allegations by the gadfly politician Subramanian Swamy of corrupt pressure brought to bear on the company's previous owners. *Organiser*, 20 May 2012, p. 2. See also Subramanian Swamy, *2G Spectrum Scam* (New Delhi: Har-Anand Publications, 2012).
76. Interview, T. V. Ramachandran, with R. Jeffrey, New Delhi, 24 February 2012.

77. Kamal Sharma et al., 'BSNL: Ringing Change in Rural India', in Arindam Mukherji (ed.), *The Icfai University Press on Mobile Service Providers* (Hyderabad: Icfai University Press, 2009), p. 223.
78. www.bsnl.co.in/about.htm (accessed 10 March 2011).
79. Doron's friends in Banaras were fond of the joke about BSNL. For similar remarks in the Lok Sabha on 27 July 2009, http://news.webindia123.com/news/articles/India/20090727/1305071.html (accessed on 19 March 2012).
80. *TRAI Annual Report, 2006–07*, p. 49 and *20010–11*, p. 21.
81. Richard White, *Railroaded. The Transcontinentals and the Making of Modern America* (New York: W. W. Norton, 2011), pp. xxxiv and 354–5.
82. There is debate about whether governments need interfere at all in the allocation of Radio Frequency; but since the 1920s, they have done so, and they are not likely to give up the power. See Robert Horvitz in *Financial Express*, 27 November 2011, www.financialexpress.com/printer/news/716612 (accessed on 27 November 2010).
83. White, *Railroaded*, p. 511.
84. *BusinessLine*, 11 January 2008, http://www.thehindubusinessline.com/todays-paper/article1612977.ece (accessed on 3 February 2011). *Times of India*, 25 June 2008. Paranjoy Guha Thakurta, 'Will Someone Take This Call?', *Tehelka*, 15 May 2010, http://tehelka.com/story_main44.asp?filename=Ne150510will_someone.asp (accessed on 3 February 2011).
85. The figure of Rs 176,000 crores was a top estimate, which was popular in media reports. 'A more realistic estimate', concluded *EPW*, was Rs 58,000 to Rs 66,000 crores—about US $12 billion. *EPW*, 4 December 2010, p. 8.
86. Comptroller and Auditor-General, *Performance Audit Report*.
87. Ibid., p. vii.
88. Ibid., pp. 33 and 35.
89. Ibid., pp. 28–9. www.thehindu.com/news/article889943.ece?service=mobile (accessed on 20 March 2012).
90. Ibid., p. 56.
91. 'Text of the Press Statement given by Shri Kapil Sibal on 2G Spectrum Issues', http://www.pib.nic.in/newsite/erelease.aspx (accessed 8 February 2011).
92. *Business Standard*, 3 February 2012, http://www.business-standard.com/india/news/sc-cancels-122-telecom-licences-tells-govt-to-take-policy-action/463582/ (accessed on 2 March 2012).
93. *Outlook* magazine was one of those that broke the story in its issues of 29 November 2010, pp. 32–49, and 6 December 2010, pp. 34–40. Politicians learned the dangers of mobile phones as the technology spread. The landmark Australian case was in 1987 when two politicians in the same party were overheard obscenely demeaning a colleague—http://australianpolitics.com/1987/03/23/kennett-peacock-car-phone-conversation.html (accessed 14 March 2011).

94. *Frontline*, 11 March 2011, p. 10.
95. Mittal, *India's New Entrepreneurial Classes*, p. 17.

3. MISSIONARIES OF THE MOBILE

1. Quoted in Claude S. Fischer, *America Calling. A Social History of the Telephone to 1940* (Berkeley, CA: University of California Press, 1992), p. 70.
2. *Live Mint*, 18 August 2010, epaper.livemint.com/ArticleImage.aspx?article=18_08_2010_003_003&mode=1 (accessed 14 July 2011).
3. *Statistics of India* (Mumbai: Tata Services, for relevant years).
4. TRAI, *Annual Report 2005–06*, p. 100.
5. Fisher, *America Calling*, p. 85.
6. *TRAI Annual Report, 2005–06*, pp. 47–8.
7. *Business Today*, 7 December 2003, archives.digitaltoday.in/businesstoday/20031207/books.html (accessed 15 July 2011).
8. *Business Today*, 23 May 2004, archives.digitaltoday.in/businesstoday/20040523/cover2.html (accessed 15 July 2011).
9. *Business Today*, 7 December 2003, ibid.
10. Interview, Kapil Arora, Senior Vice President and Country Head—Team Vodafone, Ogilvy and Mather, with R. Jeffrey, Mumbai, 28 March 2012.
11. *Business Today*, 23 May 2004, ibid.
12. *India Today*, 3 June 2011, indiatoday.intoday.in/site/articlePrint.jsp?aid=140263, for Rahul Singh on the book, Dhiraj Nayyar (ed.), *Dog Stories* (New Delhi: Natraj, 2011). en.wikipedia.org/wiki/Cheeka_(dog) (both accessed 15 July 2011).
13. C. A. Sanat Pyne, 'Launch of Vodafone Essar', n.d. [but 2007], www.caclubindia.com/forum/hutch-to-vodafone-business-strategy-107895.asp (accessed 14 July 2011).
14. Interview, Kapil Arora, 28 March 2012. *TRAI Annual Report, 2009–10*, pp. 27–8. Airtel was the leading provider with close to 130 million subscribers. Reliance, which offered both GSM and CDMA service, was roughly equal to Vodafone.
15. By March 2012, there had been three different Cheekas and some of the TV commercials had been filmed as far away as South Africa. Interview, Kapil Arora, 28 March 2012.
16. *India Today*, 3 June 2011, ibid. For rabies and biting dogs, *Tehelka*, 24 March 2012, p. 14.
17. *International Herald Tribune*, 8 August 2012, p. 1.
18. The calculation is based on the estimate of five persons to a household. In 2011, a similar calculation, based on a population of 1,200 million and 160 million TV households, suggest two-thirds of all Indians live in a household with a television set.
19. Prakash Tandon, *Return to Punjab* (Berkeley: University of California Press,

1981), Gurcharan Das, *India Unbound* (New York: Alfred A. Knopf, 2001) and William Mazzarella, *Shoveling Smoke: Advertising and Globalization in Contemporary India* (Durham, NC: Duke University Press, 2003) discuss complexities of advertising in India.
20. Mazzarella, *Shoveling Smoke*, p. 175 and Chapters 5 and 6.
21. Dhananjay Khanderao Keskar, 'Reliance Infocomm', in Arindam Mukherjee (ed.), *The Icfai Press on Mobile Service Providers: Perspectives and Practices* (Hyderabad: Icfai University Press, 2009), p. 200.
22. For the launch techniques of *Eenadu* and *Dainik Bhaskar*, see Jeffrey, *India's Newspaper Revolution*, pp. 68–9, 240.
23. *Outlook*, 5 May 2003, http://m.outlookindia.com/story.aspx?sid=4&aid=220032 (accessed 11 July 2012).
24. 'Vodafone's serious commitment to India', 1 June 2010, Vodafone internal document. Jeffrey is grateful to Neil Gough of Vodafone for making this available at an interview, New Delhi, 2 June 2010.
25. Interview, Anuradha Aggarwal, Senior Vice President, Consumer Insights and Communication, Vodafone India Limited, with R. Jeffrey, Mumbai, 28 March 2012.
26. Cellular Operators Association of India [hereafter COAI], *Annual Report, 2002–03*, pp. 8, 43.
27. A. Neela Radhika, and A. Mukund, 'Airtel Magic—Selling a Pre-paid Cellphone Service', ICFAI University Press, 2003, http//www.scribd.com/doc/54318936/Airtel-Magic-Selling-a-Pre-Paid-Cellphone-Service (accessed on 21 September 2012).
28. Interview, Bobby Sebastian, Circle Head, India Telecom Infra Ltd, with R. Jeffrey, Kochi, 24 November 2010.
29. 'Vodafone's serious commitment to India', 1 June 2010, Vodafone internal document.
30. The SIM cards themselves were the product of complex processes. Manufactured largely in China, they required the stamping of minute metal-alloy circuitry onto the tiny plastic base that eventually went into a mobile phone. Indian providers had to import the cards in large quantities. As late as 2010, Indian governments fretted over the fact that India was not making its own SIM cards and that this posed 'a grave security threat'. See *Economic Times*, 12 November 2010, http://m.economictimes.com/PDAET/articleshow/6910564.cms (accessed 1 August 2011).
31. McDonald, *Mahabharat in Polyester*, pp. 304–05. Reliance and Tata used CDMA (Code Division Multiple Access) technology, not the European standard GSM in which 'genuine' Indian mobile phone licensees had invested.
32. TRAI, *Annual Report, 2005–06*, pp. 36–7.
33. Cellular Operators Association of India [hereafter COAI], *Annual Report, 2002–03*, pp. 5, 9.
34. Ibid., p. 51.
35. Desai, *India's Telecommunications Industry*, p. 134.

36. *Businessline*, 15 April 2010, www.thehindubusinessline/com/2010/04/15/stories/20100415507202000.htm (accessed 16 December 2010).
37. Shailesh Shah, 'Upgrade or Perish: D. Satish Babu', *indiaretailing.com*, 19 July 2011, www.indiaretailing.com/person-of-the-week.asp (accessed 19 July 2011).
38. Interview, Ramesh Barath, Vice-President, New Business Development and Marketing, UniverCell, with R. Jeffrey, Chennai, 15 November 2010.
39. P. Madhusudhan Reddy, 'Univercell plans to come out with IPO soon', *andhrabusiness.com*, 28 June 2010, http://andhrabusiness.com/NewsDesc.aspx?NewsId=Univercell-plans-to-come-out-with-IPO-soon.html (accessed 19 July 2011). www.univercell.in/mobiles/populatestaticabout.action (accessed 19 July 2011).
40. Interview, Ramesh Barath, Vice-President, New Business Development and Marketing, UniverCell, with R. Jeffrey, Chennai, 15 November 2010.
41. Thomas H. Eriksen, *Globalization: The Key Concepts* (New York: Berg, 2007), p. 53.
42. Gerald Perschbacher, *Wheels in Motion: the American Automobile Industry's First Century* (Iola, WI: Krause Publications, 1996), p. 44. Sally H. Clarke, *Trust and Power* (New York: Cambridge University Press, 2007), p. 82.
43. Visits, R. Jeffrey, 16 and 17 March 2009, Allahabad, with Professor Badri Narayan and Professor M. Aslam. See also Robin Jeffrey and Assa Doron, 'Celling India: Exploring a Society's Embrace of the Mobile Phone', *South Asian History and Culture*, vol. 2, no. 3 (2011), pp. 397–416.
44. Interview, Ramesh Barath and Sunil Ramachandran, with R. Jeffrey, Chennai, 15 November 2010.
45. Interview, Anuradha Aggarwal, with R. Jeffrey, Mumbai, 28 March 2012.
46. Doron visited Ravi Varma's shop on various occasions from 2009–2012 and is grateful for his hospitality and co-operation.
47. Interview, Ravi Varma, with A. Doron, Banaras, 4 February 2011.
48. Craig Jeffrey, Patricia Jeffery and Roger Jeffery, *Degrees without Freedom? Education, Masculinities, and Unemployment in North India* (Stanford: Stanford University Press, 2008).
49. Nimi Rangaswamy and S. Nair, 'The Mobile Phone Store Ecology in a Mumbai Slum Community: Hybrid Networks for Enterprise', *Information Technologies & International* Development, vol. 6, no. 3 (2010), p. 52.
50. Interview, Anuradha Aggarwal, with R. Jeffrey, Mumbai, 28 March 2012.
51. Ibid.

4. MECHANICS OF THE MOBILE

1. *Hindu*, 2 February 2012, http://www.thehindu.com/news/national/article2853159.ece?homepage=true accessed 2 February 2012). The company was Uninor, a joint venture of Telenor of Norway and Unitech.
2. Sunil Mani, 'The Mobile Communications Services Industry in India: Has It

Led to India Becoming a Manufacturing Hub for Telecommunication Equipment?', *Pacific Affairs*, vol. 85, no. 3 (September 2012) pp. 511–30, quoting *Annual Survey of Industries* statistics. Mittal, *India's New Entrepreneurial Classes*, p. 17, estimated 3.5 million employed in all aspects of the telecom in 2005.
3. Perschbacher, *Wheels in Motion*, pp. 50–1.
4. Bureau of Labour Statistics, 'Automotive Industry: Employment, Earnings, and Hours', February 2012, http://www.bls.gov/iag/tgs/iagauto.htm (accessed 13 April 2012).
5. Interview, Sabyasachi Patra, head, government relations, Nokia, Sriperumbudur, with R. Jeffrey, Sriperumbudur, 15 November 2010. Patra became executive director of the peak body of India's information-technology hardware industry in December 2011. http://news.ciol.com/News/Executive-Track/News-Reports/Sabyasachi-Patra-New-executive-director-of-MAIT/157519/0/ (accessed 11 July 2012).
6. Anibel Ferus-Comelo and Paivi Poyhonen, *Phony Equality: Labour Standards of Mobile Phone Manufacturers in India* (Helsinki: Finnwatch, Cividep and SOMO, 2011), pp. 4, 9, http://cividep.org/wp-content/uploads/Phony_Equality.pdf (accessed 6 February 2012).
7. Ibid., p. 19.
8. Ibid., pp. 20–1.
9. Ibid., p. 18.
10. Ibid., p. 37.
11. Ibid., p. 23.
12. Ibid., p. 23.
13. Ibid., p. 28.
14. Ibid., p. 34.
15. Ibid., pp. 40, 42. *Frontline*, 19 November 2010, pp. 37–40.
16. http://www.youtube.com/watch?v=ncHdu6TLt-U (accessed 7 February 2012).
17. Dilip Subramanian, *Telecommunications Industry in India: State, Business and Labour in a Global Economy* (New Delhi: Social Science Press, 2010), p. 249.
18. Ibid., p. 373.
19. Interview, Sudhir Kumar Mehra, General Manager, ITI Limited, with A. Doron, Raebareli, 10 February 2011.
20. NGOs increasingly report deplorable working conditions in mobile phone factories in Asia, most of which employ young, rural women. For employment conditions in Nokia factories in China, see http://www.sask.fi/english/magazine/makeitfair/ (accessed 15 April 2012)
21. http://industowers.com/ (accessed 8 February 2012).
22. http://industowers.com/vision_mission.php (accessed 8 February 2012).
23. Interview, Bobby Sebastian with R. Jeffrey, Kochi, 24 November 2010.
24. *Hindu*, 6 March 2012, http://www.thehindu.com/news/cities/chennai/article2964946.ece (accessed 22 June 2012). We touch on polluting effects of diesel emissions in the Conclusion.

25. *Hindustan* (Lucknow), 4 June 2010, p. 6.
26. The term *mistrii* means 'skilled artisan' or 'master mason', and seems to have originated from either the Portuguese word 'mestre' or the Dutch word 'mester' (both meaning master). In north India the title *mistrii* is usually associated with builders, constructions workers, carpenters and stonemasons. See M.S.R. Dalgado, *Portuguese Vocables in Asiatic Languages* (Delhi: Asian Education Services, 1988), p. 228.
27. Nimmi Rangaswamy and Sumitra Nair, 'The Mobile Phone Store Ecology in a Mumbai Slum Community: Hybrid Networks for Enterprise', *Information Technologies and International Development*, vol. 6, no. 3 (2010), pp. 51–65.
28. See Ravi Sundaram's insightful analysis of Delhi's media-scapes and grey markets, *Pirate Modernity: Delhi's Media Urbansim* (New Delhi: Routledge, 2010).
29. Daal Mandi refers to any large market in UP which deals with a variety of goods, traditionally food items, hence the name Dal (lentil) Mandi (market). Many towns in UP have a Daal Mandi where both second-hand items and semi-illicit commodities are sold.
30. Interview, Sabyasachi Patra, head, government relations, Nokia, Sriperumbudur, with R. Jeffrey, Sriperumbudur, 15 November 2010.
31. Arvind Rajagopal, 'The Violence of Commodity Aesthetics: Hawkers, Demolition Raids, and a New Regime of Consumption', *Social Text*, vol. 19, no. 3 (2001), pp. 91–113.
32. E. B. White, 'Farewell, My Lovely', *New Yorker*, 16 May 1936, http://www.wesjones.com/white1.htm (accessed 9 February 2012).
33. Ibid.
34. Nimmi Rangaswamy and Sumitra Nair, 'The Mobile Phone Store', p. 62.
35. Leela Fernandes examines training and educational institutes catering for the aspiring middle classes in *India's New Middle Class* (Minneapolis: University of Minnesota Press, 2006), pp. 98–100.
36. The International Mobile Equipment Identity (IMEI) number is unique to each phone and functions like an electronic fingerprint, transmitted every time a handset is used and revealing the identity of the handset. Many China Mobiles and smuggled phones had fake IMEI numbers until the Indian government banned these phones from networks for security reasons late in 2009.
37. David Edgerton, *The Shock of the Old: Technology and Global History since 1900* (New York: Oxford University Press, 2007), p. 69.
38. For example, Eriksen, *Globalization*.
39. Ira Raja, 'Rethinking Relationality in the Context of Adult Mother-Daughter Caregiving in Indian Fiction', *Journal of Aging, Humanities and the Arts*, vol. 3 (2009), pp. 25–37.
40. The film director, Shekhar Kapur, had similar anxieties when his Blackberry failed, but once he surrendered his mobile to the street-side repair shop in Mumbai, he was enlightened, as he wrote on his blog, http://shek-

harkapur.com/blog/2010/07/a-blackberry-addict-discovers-grassroots-enterprise-in-india/ (accessed 26 July 2012).
41. For discussion of the seeming dichotomy of the formal/informal divide, see Assa Doron, 'Consumption, technology and adaptation: care and repair economies of mobile phones in north India', *Pacific Affairs*, vol. 85, no. 3 (September 2012), pp. 563–86.
42. Michael Herzfeld, *The Body Impolitic: Artisans and Artifice in the Global Hierarchy of Value* (Chicago, IL: University of Chicago, 2003).

5. FOR BUSINESS

1. For example, A. Kumar, A. Tewari, G. Shroff, D. Chittamuru, M. Kam, and J. Canny, 'An Exploratory Study of Unsupervised Mobile Learning in Rural India', paper presented at CHI 2010 conference, Atlanta, 10–15 April 2010, http://www.cs.cmu.edu/~anujk1/CHI2010.pdf (accessed 28 February 2012). D. Raha and S. Cohn-Sfetcu,'Turning the Cellphone into an Antipoverty Vaccine', *Journal of Communications*, vol. 4, no. 3 (2009), pp. 203–210.
2. Peter D. O'Neill, 'The "Poor Man's Mobile Telephone": Access versus Possession to Control the Information Gap in India', *Contemporary South Asia*, vol. 12, no. 1 (2003), p. 98.
3. Rajat Kathuria, Mahesh Uppal and Mamta [sic], 'An Econometric Analysis of the Impact of the Mobile', in *India: the Impact of Mobile Phones* (New Delhi: Vodafone, 2009), pp. 8, 14.
4. Surabhi Mittal, Sanjay Gandhi and Gaurav Tripathi, 'The Impact of Mobile Phones on Agricultural Productivity", in Kathuria *et al.*, *India: the Impact of Mobile Phones*, p. 32.
5. Interviews with A. Doron, Panaji, Goa, February 2010.
6. Mahesh Uppal and Rajat Kathuria, 'The Impact of Mobiles in the SME Sector', in Kathuria *et al.*, *India: the Impact of Mobile Phones*, pp. 54–60.
7. There are, of course, forgeries and illicit SIMs circulating. For example, http://janamejayaneconomics.wordpress.com/2012/06/04/cell-phone-menace-sim-terror/ (accessed 26 June 2012). For crime, see Chapter 8.
8. T. T. Sreekumar, 'Mobile Phones and the Cultural Ecology of Fishing', *Information Society*, vol. 27, no. 3 (2011), p. 174.
9. *New York Times*, 4 August 2001, www.nytimes.com/2001/08/04/technology/04PHON.html? (accessed 1 September 2011).
10. *Washington Post*, 15 October 2006, http://www.washingtonpost.com/wp-dyn/content/article/2006/10/14/AR2006101400342.html (accessed 1 September 2011). The fishermen were still getting international recognition six years later. See Shashi Tharoor, 'The cell phone revolution: Mobile phones have empowered India's underclass', *Pittsburgh Post-Gazette*, 13 May 2012, http://www.post-gazette.com/stories/opinion/perspectives/the-cell-phone-revolution-mobile-phones-have-empowered-indias-underclass-635700/ (accessed 24 June 2012).

11. Robert Jensen, 'The Digital Provide: Information (Technology), Market Performance and Welfare in the South Indian Fisheries Sector', *Quarterly Journal of Economics*, vol. 122, no. 3 (2007), p. 883.
12. Ibid., p. 879.
13. Ibid., p. 883.
14. Reuben Abraham, 'Mobile Phones and Economic Development: Evidence from the Fishing Industry in India', *Information Technology and International Development*, vol. 4, no. 1 (2007), p. 9.
15. C. K. Prahalad, *The Fortune at the Bottom of the Pyramid* (Upper Saddle River, NJ: Wharton School Publishing, 2005), pp. 15–16.
16. This was a more modest estimate than the 1.2 per cent increase to State Domestic Product put forward in *India: the Impact of Mobile Phones*, which had been sponsored by Vodafone. (See Note 3 of this chapter).
17. *Economist*, 10 May 2007, www.economist.com/node/9149142 (accessed 1 September 2011).
18. *Washington Post*, 15 October 2006, http://www.washingtonpost.com/wp-dyn/content/article/2006/10/14/AR2006101400342.html (accessed 1 September 2011).
19. Sanjay Gandhi, Surabhi Mittal and Gaurav Tripathi, 'The Impact of Mobiles on Agricultural Productivity', in *India: the Impact of Mobile Phones*, Policy Papers Series No. 9 (New Delhi: Vodafone Group, 2009), p. 29. Fishermen in Goa told Doron similar stories about lives saved by mobile phones during freak storms in February 2010.
20. Thakazhi Sivasankara Pillai, *Chemmeen* (Bombay: Jaico, 1964), p. 79.
21. *Washington Post*, 15 October 2006, http://www.washingtonpost.com/wp-dyn/content/article/2006/10/14/AR2006101400342.html (accessed 1 September 2011).
22. Sreekumar, 'Mobile Phones and the Cultural Ecology of Fishing', p. 174.
23. Horst and Miller coined the term 'expansive realization' to describe the process by which the mobile phone allows old practices and activities to spread and be elaborated on: 'technology is used initially with reference to desires that are historically well established, but remain unfulfilled because of the limitations of previous technologies'. Horst and Miller, *Cell Phone*, p. 6.
24. Sreekumar, 'Mobile Phones and the Cultural Ecology of Fishing', p. 175.
25. Abraham, 'Mobile Phones', p. 13.
26. Jonathan Donner and Marcela X. Escobari, 'A Review of Evidence on Mobile Use by Micro and Small Enterprises in Developing Countries', *Journal of International Development*, vol. 22 (2010), p. 651.
27. Horst and Miller, 'From Kinship to Link-up: Cell Phones and Social Networking in Jamaica', p. 761.
28. Sreekumar, 'Mobile Phones and the Cultural Ecology of Fishing', p. 175.
29. Donner and Escobari, 'A Review', p. 651.

30. O'Neill, 'The "Poor Man's Mobile Telephone"', p. 98, takes the optimistic view that 'direct information for marketing can eliminate the urban, middlemen market-makers'.
31. White, *Railroaded*, p. 289.
32. Donner and Escobari, 'A Review', p. 651.
33. Interview, Abhishek Sinha, with R. Jeffrey, New Delhi, 2 June 2010.
34. Sam Pitroda and Mehul Desai, *The March of Mobile Money: the Future of Lifestyle Management* (New Delhi: Collins Business, 2010). Sonia Kolesnikov-Jessop, 'Mobile wallets gains currency', *International Herald Tribune*, 6 September 2011, p. I.
35. Interview Abhishek Sinha, with A. Doron, New Delhi, 1 February 2011.
36. Interview, Anupam Varghese, with A. Doron, New Delhi, 1 February 2011.
37. Anindita Adhikari and Kartika Bhatia, 'NREGA Wage Payments: Can We Bank on the Banks', *EPW*, 2 January 2011, pp. 35, 37. Another survey noted that when a cash point was more than fifteen minutes away (one kilometre's walk), its usefulness to a potential customer dropped dramatically. 'Capturing the promise of mobile banking in emerging markets', *McKinsey Quarterly*, February 2010, p. 8.
38. Stephen Leacock, 'My Financial Career', in *Literary Lapses*, many editions since 1899.
39. Adhikari and Bhatia, 'NREGA Wage Payments', p. 35.
40. 900 million if one accepts the raw figures; more than 600 million if one discounts 30 per cent for 'churn'—the lapsing of connections as people acquire new phones or new service providers. In 2006, India had only 71,000 bank branches. Pankaj Kumar and Romesh Golait, 'Bank Penetration and SHG-Bank Linkage Programme: a Critique', *Reserve Bank of India Occasional Papers*, vol. 29, no. 3 (2009), pp. 120–1.
41. Interview, Anupam Varghese, with A. Doron, New Delhi, 1 February 2011.
42. Charles Thomson, manager, customer relations, EKO, email to R. Jeffrey, 26 September 2011.
43. E. C. Thompson, 'Mobile Phones, Communities and Social Networks among Foreign Workers in Singapore', *Global Networks*, vol. 9, no. 3 (2009), pp. 359–80.
44. N. Hughes and S. Lonie, 'M-PESA: Mobile Money for the "Unbanked": Turning Cellphones into 24-Hour Tellers in Kenya', *Innovations* 2, no. 1–2 (2007). pp. 63–81.
45. [Lalitha Iyer], *Documentation on Empowering ASHAs—Mobile Money Trasfer, Distt. Shiekpura* [sic], *BIHAR* (New Delhi: Norway India Partnership Initiative Secretariat, 2010), p. 3.
46. Ibid., p. 9.
47. Ibid., p. 7.
48. Ibid., p. 8.
49. A woman's account balance showed on her mobile phone. A shopkeeper-

agent of EKO could, in theory, demand a bribe before paying funds to an EKO account holder, but a local shopkeeper was more vulnerable to local retribution and boycott than a government servant or bank clerk.

50. Nokia abandoned its mobile-banking attempt in India in March 2012. *The Register*, 12 March 2012, http://www.theregister.co.uk/2012/03/12/nokia_money_gone/ (accessed 14 April 2012).
51. Thomas K. Thomas, *BusinessLine*, 24 January 2011, www.thehindubusinessline.com/features/eworld/article1118867.ece?css=print (accessed 31 October 2011).
52. Interview, Abhishek Sinha, with R. Jeffrey, New Delhi, 2 June 2010.
53. Thomas K. Thomas, *BusinessLine*, 24 January 2011.
54. For a detailed account of the river economy and passenger distribution system in Banaras, see Assa Doron, *Caste, Occupation and Politics on the Ganges: Passages of Resistance* (Farnham: Ashgate, 2008).
55. Agar, *Constant Touch*, p. 5.
56. E. P. Thompson, 'Time, Work-Discipline, and Industrial Capitalism', *Past and Present*, vol. 38 (1967), pp. 56–97.
57. A. Gupta, 'The Reincarnation of Souls and the Rebirth of Commodities: Representations of Time in "East" and "West"', *Cultural Critique*, vol. 22 (1992), pp. 187–211.
58. Rajesh Veeraraghavan, Naga Yasodhar and Kentaro Toyama, 'Warana Unwired: Replacing PCs with Mobile Phones in a Rural Sugarcane Cooperative', *Information Technologies and International Development*, vol. 5, no. 1 (Spring 2009), p. 84.
59. Maharashtrian rats were not alone in enjoying a nice cable dinner. In the US, sharp-toothed Brooklyn squirrels brought author Andrew Blum's telecom connections crashing down. Blum, *Tubes: a Journey to the Center of the Internet* (New York: Ecco, 2012), pp. 1–2, 264.
60. Veeraraghavan *et al.*, 'Warana Unwired', pp. 92–3.
61. Ibid., p. 90.
62. Fischer, *America Calling*, p. 93.
63. Surabhi Mittal, Sanjay Gandhi and Gaurav Tripathi, *Socio-Economic Impact of Mobile Phones on Indian Agriculture*, Working Paper No. 246 (New Delhi: Indian Council for Research on International Economic Relations, 2010), p. 18.
64. Uppal and Kathuria, 'Impact', p. 54.
65. O'Neill, 'The "Poor Man's Mobile Telephone"', p. 98.
66. Mittal *et al.*, *Socio-Economic Impact*, p. 4.
67. Ibid., p. 14.
68. *Hindustan Times*, 16 October 2011, www.hindustantimes.com/StoryPage/Print/758095.aspx (accessed 11 November 2011).
69. IFFCO Kisan Sanchar Limited, 'Performance for the Financial Year Apr 10—Mar 2011', www.iksl.in (accessed 11 November 2011). Prince Mathews

Thomas, 'A Rich Harvest: the Win-Win Initiative to Help Farmers', *Forbes India*, 18 February 2011, http://business.in.com/article/work-in-progress/a-rich-harvest-the-winwin-initiative-to-help-farmers/22412/0 (accessed 11 November 2011).
70. *Indian Express*, 4 September 2011, www.indianexpress.com/story-print/841426/ (accessed 11 November 2011).
71. IFFCO Kisan Sanchar Limited, 'IFFCO', www.iksl.in (accessed 11 November 2011).
72. *DNA*, 21 December 2010, http://www.dnaindia.com/print710.php?cid=1483995 (accessed 11 November 2011).
73. Nokia too began offering mobile phone service to farmers in 2009, with free trial periods as part of a package to increase the penetrations of its brand into rural areas. The service was titled 'Nokia Life Tools'; see http://sloanreview.mit.edu/improvisations/2012/02/28/information-equals-power-nokias-sms-services-for-farmers/#.T4tg-3J2qok (accessed 15 April 2012)
74. Thomas, 'A Rich Harvest'.
75. P. F. Lazarsfeld, B. Berelson, and H. Gaudet, *The People's Choice: How the Voter Makes Up His*[sic] *Mind in a Presidential Campaign* (New York: Columbia University Press, 1944).
76. Thomas, 'A Rich Harvest'.
77. Fischer, *America Calling*, pp. 84–5.

6. FOR POLITICS

1. Vicente L. Rafael, 'The Cell Phone and the Crowd: Messianic Politics in the Contemporary Philippines', *Public Culture*, vol. 15, no. 3 (2003), pp. 399–425. Robert Jensen, 'The Digital Provide', *Quarterly Journal of Economics*, vol. 22, no. 3 (August 2007), pp. 879–924.
2. Rafael, 'Cell Phone', p. 400.
3. Rheingold, *Smart Mobs*, p. 20.
4. Ibid., pp. xii-xiii.
5. Ibid., p. 168.
6. Manuel Castells, Mireia Fernandez-Ardevol, Jack Linchuan Qiu and Araba Sey, *Mobile Communication and Society. A Global Perspective* (Cambridge, MA: MIT Press, 2007), pp. 196–7. Defeated in a later election, Roh Moo-Hyun died in 2009, apparently a suicide.
7. Ibid., pp. 201–02.
8. Ibid., p. 188.
9. Ibid., p. 211.
10. Ibid., p. xxi.
11. Evgeny Morozov, *The Net Delusion* (London: Allen Lane, 2011), p. xiii.
12. Ibid., p. 320.
13. Vivek Kumar, *India's Roaring Revolution: Dalit Assertion and New Horizons* (Delhi: Gagandeep Publications, 2006), p. 53.

14. Ajoy Bose, *Behenji: a Political Biography of Mayawati* (New Delhi: Penguin, 2008), pp. 30–1.
15. Bose, *Behenji*, pp. 28–40. There is a more detailed account of the 2007 election episode in Robin Jeffrey and Assa Doron, 'Mobile-izing: Democracy, Organization and India's First "Mass Mobile Phone" Elections', *Journal of Asian Studies*, vol. 71, no. 1 (2012), pp. 63–80
16. Quoted in Bose, *Behenji*, p. 59 and see also p. 57.
17. Barbara R. Joshi, 'Recent Developments in Inter-Regional Mobilization of Dalit Protest in India', *South Asia Bulletin*, vol. 7 (1987), p. 86.
18. BAMCEF bulletin, 1974, quoted in Bose, *Behenji*, p. 34.
19. Akhilesh Suman, *Pioneer*, 12 May 2007, http://www.dailypioneer.com/archives2/default12.asp?...hy_path_it=D%3A%5Cdailypioneer%Carchives2%5Cmay1207 (accessed 22 May 2007). M. Hasan and Chandrakant Naidu, *Hindustan Times*, 11May 2007, http://www.hindustantimes.com/StoryPage/Print/221841.aspx (accessed 27 July 2010).
20. David Plouffe, *The Audacity to Win* (New York: Viking, 2010), pp. 21, 36.
21. Barbara R. Joshi, 'Scheduled Caste Voters: New Data, New Questions', *EPW*, 15 August 1981, p. 1359. Sudha Pai, *Dalit Assertion and the Unfinished Democratic Revolution: the Bahujan Samaj Party in Uttar Pradesh* (New Delhi: Sage, 2002), pp. 37, 78.
22. *Gautama Dharmasutra*, 12.1, in Patrick Olivelle (ed. and trans.), *Dharmasutra. The Law Codes of Ancient India* (Oxford: Oxford University Press, 1999) p. 98.
23. 'Uttar Pradesh. 'Data Highlights: The Scheduled Castes. Census of India 2001', http://censusindia.gov.in/Tables_Published/SCST/dh_sc_up.pdf (accessed 15 July 2010). Kanchan Chandra, *Why Ethnic Parties Succeed: Patronage and Ethnic Head Counts in India* (New York: Cambridge University Press, 2004), p. 173.
24. Mahesh Rangarajan in *Hindustan*, 27 April 2007.
25. Interview with A. Doron, Lucknow, 7 June 2010. Names of the interviewees are available, but we preserve their anonymity here.
26. Ibid.
27. The works of Sudha Pai, *Dalit Assertion and the Unfinished Revolution*, Christoffe Jaffrelot, *India's Silent Revolution* (London: C. Hurst, 2003), Vivek Kumar, *India's Roaring Revolution* (New Delhi: Gagandeep, 2006) and Badri Narayan, *The Making of the Dalit Public in North India* (New Delhi: Oxford University Press, 2011) document the BAMCEF and BSP stories. Sunetra Choudhury, *Braking News* (Gurgaon: Hachette India, 2010), pp. 111–14, for the surprise encounters of a New Delhi TV journalist with mobile phones while covering the 2009 national elections in UP.
28. Telecom Regulatory Authority of India (TRAI), *Annual Report, 2005–06*, Figure 2.2, p. 100.
29. In 2005, minimum wages for agricultural labourers under the National

Rural Employment Guarantee Scheme (NREGA) varied from state to state. The low end was Rs 60 a day. http://nrega.nic.in/wages.pdf (accessed 11 August 2010). In 2010, the cost of a cell phone was as low as Rs 850 for a basic Tata/Reliance CDMA model, and illegal China mobiles could be purchased for as little as Rs 500.

30. Plouffe, *Audacity*, p. 378: 'we put a huge premium on ... the power of human beings' talking to human beings, online, on the phone, and at the door'.
31. TRAI, *Annual Report, 2005–06*, Table 1.3, p. 40 and TRAI, *Annual Report, 2006–7*, Table 1.3, p. 41.
32. Rural phone penetration in UP was estimated to have reached 10 phones per 100 people in 2009. Department of Telecommunications, *Annual Report, 2008–09* (New Delhi: Dept of Telecommunications, Ministry of Communications and IT, 2009), p. 129.
33. Bose, *Behenji*, p. 175.
34. Ibid., p. 176.
35. Raghav Sharma [pseudonym], 'Magic realism of *Mayajaal*', *Hard News*, May 2007, pp. 17–19, http://www.hardnewsmedia.com/2007/10/1287 (accessed 21 July 2010).
36. Mahesh Rangarajan, 'Why Mayawati Matters', *Seminar*, No. 581 (January 2008).
37. Interview, Anil Kumar, with A. Doron, Lucknow, 8 June 2010.
38. 'INDIA-Uttar Pradesh State Roads Project', Project ID P067606 (New Delhi: World Bank, 2002), p. 2, http://www-wds.worldbank.org/external/default/main?pagePK=64193027&piPK=64187937&theSitePK=523679&menuPK=64187510&searchMenuPK=64187283&siteName=WDS&entityID=000094946_00060105320186 (accessed 22 July 2010).
39. Robin Jeffrey, '[Not] Being There: Dalits and India's Newspapers', *South Asia*, vol. 24, no. 2 (December 2001), pp. 225–38.
40. Eva-Maria Hardtmann, *The Dalit Movement in India* (New Delhi: Oxford University Press, 2009), p. 3. Bose, *Behenji*, pp. 110–11.
41. Bose, *Behenji*, p. 111.
42. Interview, Satish Mishra, with R. Jeffrey, Lucknow, 7 June 2010.
43. Interview, V. Thiruppugazh, Commissioner of Information, Government of Gujarat, with R. Jeffrey, New Delhi, 8 June 2010.
44. These concerns about social justice and caste discrimination were picked up in the popular advertising campaign by the 'Idea' cellular group. For example, see http://www.youtube.com/watch?v=STZAcD2R6YI (accessed 15 April 2012).
45. *Indian Express*, 25 May 2009, http://www.indianexpress.com/storyrint/465258 (accessed 23 July 2010).
46. Election Commission of India, http://eci.nic.in/eci_main/archiveofge2009/Stats/VOLI/25_ConstituencyWiseDetailedResult.pdf (accessed 23 July 2010).

47. Interview, P. L. Punia, with R. Jeffrey, Lucknow, 5 June 2010.
48. Ibid.
49. Ibid.
50. Ibid.
51. *India Today*, 12 March 2012, cover.
52. *EPW*, 7 April 2012, pp. 80–6, for detailed statistics of the 2012 UP elections.
53. Marion Walton and Jonathan Donner, 'Red-Write-Erase: Mobile-Mediated Publics in South Africa's 2009 Elections', p. 7, paper presented at the International Conference on Mobile Communication and Social Policy, Rutgers University, New Brunswick, NJ, 9–11 October 2009, http://research.microsoft.com/en-us/people/jdonner (accessed July 2010).
54. Ibid., p. 8.
55. Rodney King died in 2012 at 47. *International Herald Tribune*, 19 June 2012, p. 6. A CNN documentary shows the footage, http://www.youtube.com/watch?v=tWhYmb1sANM (accessed 25 June 2012).
56. *Financial Times*, 8 December 2008, http://www.ft.com/intl/cms/s/0/04a981ce-c553-11dd-b516-000077b07658.html#axzz1f3yKKyHv (accessed 29 November 2011). The presidential contest took a run-off to elect Atta-Mills. Though close, the result was accepted because the election was seen to have been fair and free.
57. *International Herald Tribune*, 26–27 November 2011, pp. 1, 4.
58. http://bambuser.com/ (accessed 29 November 2011). If they wanted to be ideally equipped, they could buy for less than US $200 a Looxsee, a tiny camera with 10-hours of capacity that attached to one's ear, hat or garment. http://looxcie.com/ (accessed 29 November 2011). For video and Bambuser in the 'Arab spring', *International Herald Tribune*, 21 February 2011, p. 18.
59. Quoted by Clive Thompson, 'Establishing Rules in the Videocam Age', *Wired*, 28 June 2011, http://www.wired.com/magazine/2011/06/st_thompson_videomonitoaring/ (accessed 29 November 2011).
60. Shubhranshu Choudhary, interview with Sinderpal Singh, ISAS Video Insights, 17 February 2011, http://nuscast.nus.edu.sg/PublicEvents/1/MODVideoOnly.aspx?KEY=d5486bd9-d7c2-4315-a3f9-33ce2bc91bd6 (accessed 30 November 2011) and refer to http://cgnetswara.org/about.html (accessed 30 November 2011).
61. See *More than Maoism*, eds Robin Jeffrey, Ronojoy Sen and Pratima Singh (New Delhi: Manohar for the Institute of South Asian Studies, Singapore, 2012), for chapters on the 'Maoist movement' and tribal people.
62. 'Chhattisgarh bans cell phone in schools', *iGovernment*, 20 April 2010, http://www.igovernment.in/site/chhattisgarh-bans-cell-phone-schools-37399 (accessed 30 November 2011). 'Chhattisgarh to remove illegal cell phone towers', *Siasat*, 22 March 2010, http://www.siasat.com/english/news/chhattisgarh-remove-illegal-cell-phone-towers (accessed 30 November 2011).

NOTES pp. [161–167]

63. See Anoop Saha, 'Cellphones as a Tool for Democracy: The Example of CGNet Swara', *EPW*, 14 April 2012, pp. 23–26.
64. Smita Choudhary, 'Speak Up a Revolution', manuscript, sent to R. Jeffrey by email, 5 April 2012.
65. Shubhranshu Choudhary, interview with Sinderpal Singh, 17 February 2011.
66. Ibid.
67. *Frontline*, 12 August 2011, p. 90. 'The Baigas were given their payment after a week of the story [going] out but there was no action against the BDO [Block Development Officer]', Choudhary later reported. Email to R. Jeffrey, 30 November 2011.
68. David Singh Grewal, *Network Power: The Social Dynamics of Globalization* (New Haven: Yale University Press, 2008), pp. 6–10.
69. Grewal chooses a few dominant technologies and the components of the World Trade Organization as examples of how 'network power' arises and operates. Grewal, *Network Power*, especially Chapters 7 and 8.
70. Lee Rainie and Barry Wellman, *Networked: the New Social Operating System* (Cambridge, MA: MIT Press, 2012), pp. 285–8, 291–6. They offered two future-gazing scenarios, one cheery, the other bleak.
71. Ibid., p. 294.

7. FOR WOMEN AND HOUSEHOLDS

1. *Times of India*, 5 June 2011, http://articles.timesofindia.indiatimes.com/2011–06–05/india/29622994_1_boyfriend-cellphone-police (accessed 12 January 2012).
2. *Times of India*, 3 June 2011, http://articles.timesofindia.indiatimes.com/2011–06–03/patna/29616827_1_warisaliganj-nawada-boyfriend (accessed 12 January 2012).
3. Ibid.
4. *India Today*, 14 October 2002, http://archives.digitaltoday.in/indiatoday/20021014/cover.html (accessed 16 January 2012).
5. Jonathan Donner, Nimmi Rangaswamy, Molly Steenson and Carolyn Weil, '"Express Yourself" and "Stay Together": the middle-class Indian family', in Katz (ed.), *Handbook of Mobile Communication Studies*, pp. 325–37.
6. *Times of India*, 19 June 2011, http://articles.timesofindia.indiatimes.com/2011–06–19/india/29676266_1_mobile-phones-bans-unmarried-girls (accessed 12 January 2012).
7. *Times of India*, 23 November 2010, http://articles.timesofindia.indiatimes.com/2010–11–23/india/28235199_1_mobile-phones-panchayat-gotra-marriages (accessed 12 January 2012). See also *Hindu*, 13 July 2012, for a similar proclamation by village elders in western UP. http://www.thehindu.com/news/national/article3632680.ece (accessed 13 July 2012).

8. *Tehelka*, 31 March 2012, p. 36.
9. Anneryan Heatwole, 'Deconstructing Mobiles: Myths and Realities about Women and Mobile Phones', 16 October 2009, *Mobile Active.org*, http://mobileactive.org/deconstructing-mobiles-women-and-mobiles (accessed 9 March 2010).
10. Marvin, *When Old Technologies Were New*, p. 23, quoted in Michele Martin, '*Hello, Central*': *Gender, Technology and Culture in the Formation of Telephone Systems* (Quebec: McGill-Queen's University Press, 1991), p. 140.
11. Fischer, *America Calling*, p. 78.
12. Diane, Z. Umble, 'The Amish and the Telephone: Resistance and Reconstruction', in Roger Silverstone and Eric Hirsch (eds), *Consuming Technologies: Media and Information in Domestic Spaces* (London: Routledge, 1992), pp. 171–81.
13. Hiyam Omari-Hijazi and Rivka Ribak, 'Playing with Fire: On the Domestication of the Mobile Phone among Palestinian Teenage Girls in Israel', *Information, Communication and Society*, vol. 11, no. 2 (2008), pp. 149–66.
14. Sirpa Tenhunen, 'Mobile Technology in the Village: ICTs, Culture and Social Logistics in India', *Journal of the Royal Anthropological Society* (New Series), vol. 14 (2008), p. 524.
15. Sebastian Ureta, 'Mobilizing Poverty?: Mobile Phone Use and Everyday Spatial Mobility among Low-Income Families in Santiago, Chile', *The Information Society*, vol. 24, no. 2 (2008), p. 88.
16. Kathleen Diga, 'Mobile Cell Phones and Poverty Reduction: Technology Spending Patterns and Poverty Level Change among Households in Uganda', written April 2008 and presented at Workshop on the Role of Mobile Technologies, Sao Paolo, Brazil, 2–3 June 2008, http://www.w3.org/2008/02/MS4D_WS/papers/position_paper-diga-2008pdf.pdf (accessed 13 January 2012).
17. Cara Wallis, 'The Traditional Meets the Technological: Mobile Navigations of Desire and Intimacy', in S. H. Donald, T. D. Anderson and D. Spry (eds), *Youth, Society, and Mobile Media in Asia* (London: Routledge, 2010), p. 59.
18. Steven Derné, 'Hindu Man Talk About Controlling Women: Cultural Ideas as the Tool of the Powerful', *Sociological Perspectives*, vol. 37, no. 2 (1994), pp. 203–27. Chanasai Tiengtrakul, 'Home: Banarasi Women and Perceptions of the Domestic Domain', in Lina Fruzzetti and Sirpa Tenhunen (eds) *Culture, Power, and Agency: Gender in Indian Ethnography* (Kolkata: Stree, 2006), pp. 22–51.
19. Sara Dickey, 'Permeable Homes: Domestic Service, Household Space, and the Vulnerability of Class Boundaries in Urban India', *American Ethnologist*, vol. 27, no. 2 (2000), p. 470.
20. Such spatial classification and gender relations are not static and have changed overtime. For example, Judith Walsh, *Domesticity in Colonial India:*

What Women Learned When Men Gave Them Advice (Oxford: Rowman and Littlefield, 2004).
21. Gloria G. Raheja and Ann G. Gold, *Listen to the Heron's Words: Reimagining Gender and Kinship in North India* (Berkeley: University of California Press, 1994).
22. Cf. Omari-Hijazi and Ribak, 'Playing with Fire', p. 153.
23. *Hindu*, 13 July 2012, http://www.thehindu.com/news/national/article3632680.ece (accessed 13 July 2012).
24. Sara Dickey, 'Permeable Homes', p. 474.
25. Sarah Lamb, 'The Making and Unmaking of Persons: Notes on Aging and Gender in North India', *Ethos*, vol. 25, no. 3 (1997), pp. 279–302.
26. Lamb, 'The Making and Unmaking', pp. 289–90
27. Srimati Basu, *She Comes to Take Her Rights: Indian Women, Property, and Propriety* (Albany: SUNY Press, 1999).
28. Patricia Uberoi, *Freedom and Destiny: Gender, Family, and Popular Culture in India* (New Delhi: Oxford University Press, 2006), p. 31.
29. Ibid.
30. Katherine Boo wrote about similar telephonic assignations in *Behind the Beautiful Forevers* (London: Hamish Hamilton, 2012), p. 182.
31. Michel Foucault, *Ethics: Subjectivity and Truth*, ed. Paul Rabinow, trans. Robert Hurley and others (New York: New Press, 1997). Laura Ahearn, *Invitations to Love: Literacy, Love Letters, and Social Change in Nepal* (Ann Arbor: University of Michigan Press, 2001).
32. For detailed discussion of mobile phones as a 'technology of the self' in India, see Assa Doron 'Mobile Persons: Cell Phones, Gender and the Self in North India', *The Asia Pacific Journal of Anthropology*, vol 13, no. 4 (2012).
33. Leela Dube, 'On the Construction of Gender: Hindu Girls in Patrilineal India', *EPW*, 4 June 1988, p. 11–19.
34. See Craig Jeffrey *et al.*, *Degrees without Freedom*.
35. Uberoi, *Freedom and Destiny*, p. 252.
36. Anand Giridharadas, *India Calling* (Melbourne: Black Inc., 2011), p. 172.
37. Horst and Miller, *Cell Phone*, p. 57.
38. Ibid., p. 173, citing R. T. Smith, *The Patrilocal Family: Power, Pluralism and Politics* (New York and London: Routledge, 1996).
39. The study of mobile phone usage among Palestinian teenage girls in Israel suggests similar challenges and rigidities. Patriarchy continues to set the tone for household and gender relations, and planned marriage remains a common practice. See Omari-Hijazi and Ribak, 'Playing with Fire', p. 159.
40. Daniel Miller, *Tales from Facebook* (Cambridge: Polity, 2011), pp. 167–8. See also p. 174, for Miller's musings on what effect Facebook may have on Trinidad's 'shy East Indian girls' and the possible 'eventual extinction of this stereotype'.

8. FOR 'WRONGDOING': 'WAYWARDNESS' TO TERROR

1. How to manage, control and define what constitutes obscenity and the 'public good' is a politically charged issue. For an excellent summary of these debates in India, see Brinda Bose (ed.), *Gender and Censorship* (Delhi: Women Unlimited, 2006).
2. See Anita Gurumurthy and Nivedita Menon, 'Violence Against Women via Cyberspace', *EPW*, vol. 44, no. 40, 3 October 2009, pp. 19–21.
3. *Indian Express*, 4 February 2012, quoting Justice Altamas Kabir, www.indianexpress.com/story-print/907671 (accessed on 2 June 2012).
4. *Tehelka*, 31 March 2012, p. 36.
5. 'Wali' or male 'wala' is a common suffix to denote someone who is engaged in the business of 'x'. Because it is increasingly common to see women speaking on the phone, such a woman is referred to as a Mobile Wali.
6. Some of these can be viewed on YouTube, for example, http://www.youtube.com/watch?v=2vcmwohzFHs; and
7. See http://articles.timesofindia.indiatimes.com/2011–05–14/patna/2954 2871_1_bhojpuri-films-bhojpuri-cinema-bollywood-movies and http://www.sunday-guardian.com/artbeat/bhojpuri-cd-migrant-movies (both accessed 20 February 2012)
8. Vishal Rawlley, 'Miss Use: A Survey of Raunchy Bhojpuri Music Albums', *Tasveer Ghar: A Digital Archive of South Asian Popular Visual Culture*, http://tasveergharindia.net/cmsdesk/essay/66/index.html (accessed 10 May 2010).
9. The Bhopuri-speaking area comprises eastern UP, Bihar and Jharkhand, although with recent migration, Bhojpuri is now heard and spoken across India and beyond. Rawlley, 'Miss Use'.
10. Manuel also speaks of the parody and raunchiness characteristic of these folk tunes. Manuel, *Cassette Culture*, pp. 133, 172.
11. Compare this Bhojpuri clip with the more tame, but equally suggestive, Bollywood film song from *Haseena Maan Jaayegi*. The first words are in English: 'What is you mobile number? What is your smile number? Can we have a private party? What is your private number?', featuring Karisma Kapoor and Govinda. See http://www.youtube.com/watch?v=HhHrV5 UsQ_Q (accessed 10 May 2010).
12. The association between loose hair and sexuality has long been a theme in South Asian culture.
13. See http://www.youtube.com/watch?v=NJOlV0nhVko&feature=relmfu (accessed 20 February 2012). This song draws on a long tradition of the Bhojpuri Brahe tunes associated with different low castes, such as washermen and boatmen. See Edward O. Henry, 'Social Structure and Music: Correlating Musical Genres and Social Categories in Bhojpuri-Speaking India', *International Review of the Aesthetics and Sociology of Music*, vol. 19, no. 2 (1988), pp. 217–27.

14. There were many more film clips that played upon the 'mobile wali' theme, expressing similar desires and anxieties. For example, http://www.youtube.com/watch?v=4SlwNJbpX5k and http://www.youtube.com/watch?v=prGSJQes8_w&feature=related (accessed 10 April 2012); see also the India Ink blog in the *New York Times* discussing such songs, http://india.blogs.nytimes.com/2012/06/20/colloquial-music-spawns-a-culture-of-romance/ (accessed 14 July 2012)
15. Duggal, *Mobile Law*, pp. 237, 353, points to section 67 of the Information Technology Act, 2000, as the legislation governing electronic pornography. Possession is not a crime; publishing or transmitting is.
16. Sumit Sarkar, '"Kaliyuga", "Chakri" and "Bhakti": Ramakrishna and His Times', *EPW*, 18 July 1992, p. 1544.
17. Ibid., p 1550. See also Ratnabali Chatterjee, 'Representations of Gender in Folk Paintings of Bengal', *Social Scientist*, vol. 28, nos. 3/4 (2000), pp. 7–21.
18. The arrival of the railways in the nineteenth century also raised patriarchal concerns. Tanika Sarkar, *Hindu Wife, Hindu Nation: Community, Religion, and Cultural Nationalism* (London: Hurst, 2001), pp. 80–1. See also Christopher Pinney, 'On living in the kal(i)yug: Notes from Nagda, Madhya Pradesh', *Contributions to Indian Sociology*, vol. 33, no. 1 & 2 (1999), p. 93.
19. http://www.madaboutads.com/video/telecom/873/lava-mobiles-a10-sorry-no-change.html (accessed 15 June 2012).
20. *Times of India*, 26 April 2012, http://articles.timesofindia.indiatimes.com/2012-04-26/telecom/31409497_1_mobile-phones-mobile-handsets-touch (accessed 15 June 2012).
21. Correspondence, A. Kumar with A. Doron, 8 December 2011.
22. These were usually traveling music and dance groups that specialised in offering as part of their 'entertainment package' nude dancers for bachelor parties. Film clips were circulated by mobile phones.
23. This figure was hard to corroborate, as was the question of whether female actors earned as much as males.
24. Vinod Mehta, *Lucknow Boy* (New Delhi: Penguin Viking, 2011), p. 84
25. http://www.openthemagazine.com/article/living/porn-for-all-seasons (accessed 20 February 2012).
26. Raids by police received a boost in the late 1980s with the enactment of obscenity laws. See 'Obscenity law "a victory" for Indian women's groups', *Sydney Morning Herald*, 3 November 1987.
27. Srivastava analyses this 'minor' literature as subversive of the 'official' discourse on sexuality and 'sex education' in India. Sanjay Srivastava, *Passionate Modernity: Sexuality, Class and Consumption in India* (London: Routledge, 2007), pp. 121–46.
28. See Lotte Hoek, 'Cut-Pieces as Stag Film: Bangladeshi Pornography in Action Cinema', *Third Text*, vol. 24, no. 1 (2010), pp. 135–148; Bhrigupati Singh, 'Aadamkhor Haseena (The Man-Eating Beauty) and the Anthropol-

ogy of a Moment', *Contributions to Indian Sociology*, vol. 42, no. 2 (2008), pp. 249–79.

29. On pornography and south India, see http://www.tehelka.com/story_main5.asp?filename=Ne082804never_mind.asp (accessed 22 March 2012).

30. Duggal, *Mobile Law*, pp. 237, 353. The Indecent Representation of Women (Prohibition) Act also deals with pornography. See http://www.priyo.com/tech/2011/05/28/watching-pornography-legal-new-27397.html (accessed 22 March 2012).

31. K. Shankar, 'Sexting: A New Form of Victimless Crime?', *International Journal of Cyber Criminology*, vol. 3. no. 1 (2009), pp. 21–5.

32. Gerard Goggin, *Cell Phone Culture* (London: Routledge, 2006), pp. 107–124; Stephanie Donald, T. A. Anderson and Damien Spry (eds), *Youth, Society and Mobile Media in Asia* (London: Routledge, 2010).

33. 'Cellphone Sexcapade Fuels Scandal in India', *Los Angeles Times*, 21 December 2004, www.articles.latimes.come/print/2004/de/21/world/fg-cellsex21 (accessed 10 December 2011).

34. 'Scandal in School Shakes up Delhi', *The Telegraph*, 27 November 2004, www.telegraphindia.com/1041127/asp/frontpage/story_4055884.asp (accessed 10 December 2011).

35. 'Cellphone Sexcapade', *Los Angeles Times*, 21 December 2004, ibid.

36. *'Love Sex aur Dhoka'*, Wikipedia, www.en.wikipedia.org/wiki/Love_Sex_aur_Dhoka and 'Dev D is based on DPS MMS scandal', *Oneindia Entertainment*, 7 January 2009, www.greynium.com/mail-print/print.php (both accessed 10 December 2011).

37. Another example, widely circulated in UP, was the Saharanpur Scandal. Again, a young man filmed his girlfriend in sexually explicit positions. *Tehelka*, 5 March 2011, http://www.tehelka.com/story_main49.asp?filename=hub050311COMMENT.asp (accessed 15 April 2012).

38. Gopalan Ravindran, 'Moral Panic and Mobile Phones: The Cultural Politics of New Media Modernity in India', in Erwin Alampay (ed.), *Living the Information Society in Asia* (Singapore: Institute of Southeast Asian Studies, 2009), p. 101. *Hindu*, 8 October 2007, http://www.hindu.com/2007/10/08/stories/2007100851560300.htm (accessed 12 June 2012), for an attempt by the Karnataka government to ban mobiles in schools.

39. 'Indian Police get Mobile over Porn', *Online Asia Times*, 28 July 2005, http://www.atimes.com/atimes/South_Asia/GG28Df01.html (accessed 12 December 2011).

40. Nishant Shah. 'Subject to Technology: Internet Pornography, Cyber-Terrorism and the Indian State', *Inter Asia Cultural Studies Journal*, vol. 8, no 3 (August 2007), p. 357.

41. Shirky, *Here Comes Everybody*, p. 98–9.

42. http://www.tehelka.com/story_main49.asp?filename=hub050311COMMENT.asp (accessed 20 February 2012)

NOTES pp. [196–198]

43. *Indian Express*, 21 February 2010, http://www.indianexpress.com/news/mms-row-girl-boyfriend-commit-suicide/582429/0 (accessed 12 December 2011).
44. *Times of India*, 5 July 2011, http://timesofindia.indiatimes.com/city/delhi/Girl-raped-for-6-hrs-in-moving-car/articleshow/9105242.cms (accessed 12 December 2011).
45. *Times of India*, 3 July 2011, http://timesofindia.indiatimes.com/city/mumbai/MMS-clip-blows-lid-off-gangrape/articleshow/9081565.cms (accessed 12 December 20110).
46. http://indiatoday.intoday.in/story/pornography-on-the-internet-hits-indian-society-sunny-leone/1/174081.html (accessed 14 April 2012)
47. A similar case occurred in Indonesia in 2011 when an anti-pornography campaigner and member of the Islamic Prosperous Justice Party (PKS) was caught watching porn on his computer during a parliamentary session. See http://articles.nydailynews.com/2011-04-11/news/29426601_1_indonesian-lawmaker-nazril-ariel-irham-parliamentary-session (accessed 12 April 2012).
48. Fischer, *America Calling*, p. 143.
49. '1 in 2 Indians victim of mobile phone loss: Norton', rediff.com, 18 March 2011, http://www.rediff.com/business/slide-show/slide-show-1-tech-1-in-2-indians-victim-of-mobile-phone-loss-norton/20110318.htm (accessed 7 June 2012).
50. TRAI, 'Preliminary Consultation Paper on Mobile Phone Theft', Consultation Paper No. 2/2004, 8 January 2004, p. 7, http://www.trai.gov.in/trai/upload/ConsultationPapers/29/mobile%20theft%20rev.1.pdf (accessed 7 June 2012).
51. 'Four Arrested for "Changing" Cellphone IMEI Numbers', *Indian Express*, 31 May 2012, http://www.indianexpress.com/news/four-arrested-for-changing-cellphone-imei-numbers/956021/ (accessed 8 June 2012). Section 66c of Information Technology Act of 2008 made identity theft punishable by up to three years' imprisonment.
52. *Fonearena*, 24 December 2008, http://www.fonearena.com/blog/2262/change-or-be-gone-more-time-for-korean-handsets.html (accessed 12 June 2012).
53. Lourdes M. Portus, 'How the Urban Poor Acquire and Give Meaning to the Mobile Phone', in James E. Katz (ed.), *Handbook of Mobile Communication Studies* (Cambridge, MA: MIT Press, 2008), p. 117, writes of the Philippines that 'urban poor informants became victims of these robberies despite their owning old or obsolete mobile phone models'.
54. Ananth Majumdar, 'Mobile Phone Theft—Modus Operandi & Indian Police Stations', *My World*, 14 June 2009, http://ananthsreflections.blogspot.sg/2009/06/mobile-phone-theft-modus-operandi.html (accessed 7 June 2012).
55. *Hindustan Times*, 21 October 2011, http://www.hindustantimes.com/India-

news/NewDelhi/Mobile-phone-theft-up-by-15-this-year-Police/Article1-759743.aspx (accessed 8 June 2012).
56. '29.9 M People in India Are Cyber Crime Victims: Norton', *Computer Resellers Network*, 9 September 2011, http://www.crn.in/Security-009Sep011-29-M-People-In-India-Are-Cyber-Crime-Victims-Norton.aspx (accessed 8 June 2012).
57. http://www.madhepuratimes.com/2012/04/blog-post_7255.html (accessed 4 June 2012). Translation by A. Doron. The story originated with the 'SamidhaGroup', which described itself as 'an IT awareness centre for rural Bihar'.
58. *Indian Express*, 20 March 2012, http://articles.timesofindia.indiatimes.com/2012-01-10/india/30611486_1_question-paper-mbbs-mobile-phones (accessed 8 June 2012). http://articles.timesofindia.indiatimes.com/2012-01-10/india/30611486_1_question-paper-mbbs-mobile-phones
59. *Times of India*, 10 January 2012, http://articles.timesofindia.indiatimes.com/2012-01-10/india/30611486_1_question-paper-mbbs-mobile-phones (accessed 8 June 2012).
60. Rich Ling and Jonathan Donner, *Mobile Communication* (Cambridge: Polity, 2009), p. 209.
61. Ling and Donner, *Mobile Communication*, p. 126.
62. *Times of India*, 12 July 2011, http://articles.timesofindia.indiatimes.com/2011-07-12/lucknow/29764246_1_unnao-extra-marital-affair-sohramau (accessed 7 December 2011).
63. *Times of India*, 3 June 2011, http://articles.timesofindia.indiatimes.com/2011-06-03/patna/29616827_1_warisaliganj-nawada-boyfriend (accessed 5 April 2012).
64. *Ahmedabad Mirror*, 12 October 2011, http://www.ahmedabadmirror.com/article/48/201110122011101219172460881017711f/CBI-sensation-Burnt-Nashik-officer-was-in-cahoots-with-his-killer.html (accessed 15 December 2011). *NDTV*, 27 January 2011, http://www.ndtv.com/article/india/sonawanes-killing-top-officers-on-strike-in-maharashtra-81740 (accessed 30 January 2011).
65. Ranjani Mazumdar, 'Terrorism, Conspiracy, and Surveillance in Bombay's Urban Cinema', *Social Research*, vol. 78, no. 1 (2011), p. 166.
66. One estimate claimed US $34 billion was invested by Indian telecom companies in 2009–10 alone. *International Herald Tribune*, 17–18 July 2010, p. 9, quoting COAI sources.
67. *Outlook*, 29 November 2010, pp. 32–49.
68. He was released on bail in May 2012 after 15 months as an undertrial prisoner.
69. Siddharth Varadarajan, 'Welcome to the Matrix of the Indian State', *Hindu*, 29 November 2010, www.thehindu.com/opinion/columns/siddharth-varadarajan/article920054/ece?service=mobile (accessed 29 November 2010).

70. *Hindu* (Thiruvananthapuram), 23 November 2010, p. 10.
71. *Indian Express*, 15 August 2011, http://www.indianexpress.com/news/deadline-over-only-2-agencies-report-tapping-equipment/832052/ (accessed 11 June 2012).
72. Rudyard Kipling, 'A Code of Morals', http://www.poetryloverspage.com/poets/kipling/code_of_morals.html (accessed 11 June 2012).
73. Qiu, *Working-Class Network Society*, p. 188.
74. Morozov, *Net Delusion*, p. 5.
75. Rainie and Wellman, *Networked*, pp. 294–6.
76. *Indian Express*, 2 May 2010, http://www.indianexpress.com/news/phone-tapping-govt-looks-at-safeguards/613956/ (accessed 11 June 2012).
77. http://www.clear-trail.com/ and http://www.shoghicom.com/companyoverview.html (accessed 11 June 2012).
78. Praveen Swami, 'The government's listening to us', *Hindu*, 1 December 2011, http://www.thehindu.com/news/national/article2678501.ece (accessed 11 June 2012).
79. Praveen Swami, 'The government's listening to us', *Hindu*, 1 December 2011, http://www.thehindu.com/news/national/article2678501.ece (accessed 11 June 2012).
80. Wikipedia (2011), 'Yahya Ayyash', http://en.wikipedia.org/wiki/Yahya_Ayyashhttp://en.wikipedia.org/wiki/Yahya_Ayyash (accessed 6 December 2011).
81. Wikipedia (2011) '2004 Madrid Bombings', http://en.wikipedia.org/wiki/2004_Madrid_train_bombings#cite_note-70 (accessed 6 December 2011).
82. Manuel Castells, Mireia Fernandez-Ardevol, Jack Qui, and Araba Sey, *The Mobile Communication Society: A Cross-Cultural Analysis of Available Evidence on the Social Uses of Wireless Communication Technology* (Los Angeles: University of Southern California, 2004), pp. 217–220. http://hack.tion.free.fr/textes/MobileCommunicationSociety.pdf (accessed 5 April 2012)
83. There are many accounts on this episode, including a well-referenced version at http://en.wikipedia.org/wiki/Kargil_War (accessed 12 June 2012). B. Raman, a hawkish Indian analyst, digests the Indian view in 'Release of Kargil tape: Masterpiece or blunder?', *Rediff India Abroad*, 27 June 2007, http://www.rediff.com/news/2007/jun/27raman.htm (accessed 11 June 2012). Pakistani forces retreated across the Line of Control by the end July. In October 1999, Musharraf overthrew the government and ran Pakistan as a military dictatorship until August 2008.
84. *Frontline*, 6 December 2002, http://www.frontlineonnet.com/fl1924/stories/20021206003404700.htm (accessed 15 June 2012).
85. http://en.wikipedia.org/wiki/Muhamed_Haneef (accessed 17 June 2012).
86. For example, http://wn.com/Mumbai_Attack_Terror_Tape-phone_Conversation_Part1 (accessed 13 December 2011).
87. *New York Times*, 7 January 2009, http://www.nytimes.com/2009/01/

07/world/asia/07india.html?_r=1&scp=3&sq=India&st=cse (accessed 13 December 2011).
88. *New York Times*, 7 January 2009, http://www.nytimes.com/2009/01/07/world/asia/07india.html?_r=1&scp=3&sq=India&st=cse (accessed 13 December 2011).
89. 'Suspect Stirs Mumbai Court by Confessing', *New York Times*, 20 July 2009, www.nytimes./com/2009/07/21/world/asia/21india.html (accessed 13 December 2011).
90. *Outlook*, 25 July 2011.
91. Ujjwal Kumar Singh, *The State, Democracy and Anti-Terror Laws In India* (New Delhi: Sage, 2007), pp. 292, 325.
92. Bus conductors in the southern town of Mangalore were reportedly encouraged to send SMSes to vigilante groups of Hindu chauvinists 'when they see an inter-religious couple socialise'. *Outlook*, 13 August 2012, http://goo.gl/LMTYg (accessed 31 August 2012).
93. Agar, *Constant Touch*, p. 135.
94. *Sydney Morning Herald*, 22 January 2011, http://www.smh.com.au/technology/technology-news/porn-ban-for-phones-a-case-of-moral-panic-20110121-19zyu.html (accessed 14 June 2012).
95. McKinnon, *Consent of the Networked*, pp. 38–40.
96. Ang Peng Hwa, Shyam Tekwani and Guozhen Wang, 'Shutting Down the Mobile Phone and the Downfall of Nepalese Society', *Pacific Affairs*, vol. 85, no. 3 (September 2012), pp. 547–62. *Economist*, 10 February 2011, for Egypt and the 'Arab spring' of 2011, http://www.economist.com/node/18112043/print (accessed 12 February 2011).
97. http://ptlbindia.blogspot.com.au/2012/03/cell-phone-laws-in-india.html (accessed 13 April 2012). Duggal's door-stopping volume, *Mobile Law*, was an attempt to pull together regulations, laws and discussion about electronic transgressions.
98. *Times of India*, 23 August 2010, http://articles.timesofindia.indiatimes.com/2010-08-23/bangalore/28302684_1_cyber-crime-mobile-lab-mobile-unit (accessed 14 June 2012).

CONCLUSION: 'IT'S THE AUTONOMY, STUPID'

1. 'Afterword', in James E. Katz (ed.), *Handbook of Mobile Communication Studies* (Cambridge, MA: MIT Press, 2008), p. 448.
2. See Microsoft External Research, http://research.microsoft.com/en-us/collaboration, especially 'Health and Wellbeing', Parts Nos 098–111196–9, 098–111162 and unnumbered sheets for 2008. Jeffrey was given hard copies at Microsoft in Bengaluru in November 2010.
3. *Information and Communications for Development 2012: MaximizingMobile* (Washington, DC: World Bank, 2012), p. 57, http://www.worldbank.org/ict/IC4D2012 (accessed 31 July 2012).

4. 'Electromagnetic Fields and Public Health. Base Stations and Wireless Technologies', World Health Organization, Fact Sheet, No. 304 (May 2006), www.who.int/mediacentre/factsheets/fs 304/en/print.html (accessed on 16 September 2010).
5. *Outlook*, 14 June 2010, p. 14 was just one example.
6. 'Cell Phone Radiation Warning Ordinance Reinstated in San Francisco', Filutowski Law Firm, 21 July 2011, http://www.filutowskilaw.com/2011/07/cell-phone-warning-ordinance-reinstated-in-san-francisco/ (accessed 5 March 2012). Our italics. The lawyers appeared to be gilding the lotus.
7. *Tehelka*, 5 June 2010, p. 30.
8. Ibid., p. 31.
9. *BusinessLine*, 12 April 2006, http://www.thehindubusinessline.in/2006/04/12/stories/2006041201750400.htm (accessed on 5 March 2012).
10. 'Mobile Towers Major Air Polluters', *Times of India*, 25 May 2011, http://articles.timesofindia.indiatimes.com/2011-05-25/bangalore/29581162_1_mobile-towers-reduction-in-total-costs-solar-power (accessed 10 July 2012); and 'Telcos write to Rescos; offer to take entire wind, solar output', *Economic Times*, 8 July 2012, http://articles.economictimes.indiatimes.com/2012-07-08/news/32588711_1_telecom-towers-coai-and-association-generators (accessed 10 July 2012).
11. On the conflicts surrounding the mining of rare earths, see James. H. Smith, 'Tantalus in the Digital Age: Coltan Ore, Temporal Dispossession, and "Movement" in the Eastern Democratic Republic of Congo', *American Ethnologist*, vol. 38, no. 1 (2011), pp. 17–35.
12. http://www.toxicslink.org/art-view.php?id=134(accessed on 28 March 2012).
13. http://www.un.org/apps/news/story.asp?NewsID=33845&Cr=waste&Cr1 (accessed on 28 March 2012).
14. Garima Jain in *Tehelka*, 15 January 2011, http://www.tehelka.com/story_main48.asp?filename=hub150111WHERE_COMPUTERS.asp (accessed 15 April 2012).
15. See Frank Korom, 'On the Ethics and Aesthetics of Recycling in India', in Lance Nelson (ed.), *Purifying the Earthly Body of God: Religion and Ecology in Hindu India* (New York: New York University Press, 1998), pp. 197–223. Lucy Noris, *Recycling Indian Clothing: Global Contexts of Reuse and Value* (Bloomington: Indiana University Press, 2010). On the recycling of plastic waste, see Kavery Gill, *Of Poverty and Plastic: Scavenging and Scrap Trading Entrepreneurs in India's Urban Informal Economy* (New Delhi: Oxford University Press, 2010). Boo, *Beyond the Beautiful Forevers*, focuses on the distressing life of a boy who retrieved rubbish in Mumbai.
16. Other places outside the major cities, such as Moradabad on the Ramganga river in Uttar Pradesh, have also taken up the waste industry. The city, previously known as the capital of ornamental brassware, is said to be a centre for recycling digital waste.

17. Rich Ling, *The Mobile Connection. The Cell Phone's Impact on Society* (San Francisco: Elsevier, 2004), p. 180.
18. Ibid., p. 180.
19. Castells *et al.*, *Mobile Communication and Society*, p. 251. Barry Wellman, 'Physical Place and Cyberplace: the Rise of Personalized Networking', *International Journal of Urban and Regional Research*, vol. 25, no. 2 (June 2001), p. 238.
20. Barry Wellman, 'Little Boxes, Glocalization, and Networked Individualism', [n.d.; c. 2002], http://homes.chass.utoronto.ca/~wellman/publications/littleboxes/littlebox.PDF (accessed on 8 March 2012).
21. The Government of India set up more than thirty Telecom Enforcement, Resource and Monitoring (TERM) cells around the country. It also instructed telecom companies to install technology that would allow authorities to track city customers to within 50 metres of their location. Companies protested about the huge costs required, and compliance was sluggish. *Business Today*, 27 May 2012, p. 16. Department of Telecommunications,'Telecom Enforcement, Resource and Monitoring (TERM) Cells', n.d., http://www.dot.gov.in/vtm/vtm.htm (accessed 29 June 2012).
22. Grewal, *Network Power*, pp. 294–5.
23. Ratnakar Tripathy, 'Music Mania in Small-town Bihar: Emergence of Vernacular Identities', *EPW*, 2 June 2012, p. 65.
24. Ratnakar Tripathy, email to R. Jeffrey, 19 June 2012.
25. Tripathy, 'Music Mania', p. 61. Sale of 3,000 CDs was enough to justify production.
26. Tripathy, 'Music Mania', p. 63.
27. CNN, 'Gaming Reality', 8 August 2012, http://www.cnn.com/interactive/2012/08/tech/gaming.series/ (accessed 8 August 2012), for a series of programs on the forms and economics of electronic gaming throughout the world.
28. Craig Jeffrey *et al.*, *Degrees Without Freedom*, pp. 181–5.
29. Nimmi Rangaswamy and Edward Cutrell, 'Anthropology, Development and ICTs: Slums, Youth and the Mobile Internet in Urban India', p. 4, *ICTD2012*, 12–15 July 2012, Atlanta, GA, http://research.microsoft.com/en-us/um/people/cutrell/ICTD2012-Rangaswamy_Anthropologists_and__ICTD.pdf (accessed 8 August 2012).
30. Ibid., p. 4.
31. Ibid., p. 6.
32. Ibid., pp. 7–8.
33. CNN, 'Gaming Reality', 8 August 2012, http://www.cnn.com/interactive/2012/08/tech/gaming.series/korea.html (accessed 8 August 2012).
34. Census of India 2011. *Rural Urban Distribution of Population (Provisional Population Totals)* (New Delhi: Registrar General and Census Commissioner, 2011), p. 5. 69 per cent of the total population was classified as rural. Of rural households, 55 per cent were estimated to have mobile phones. *Times*

of India, 14 March 2012, http://epaper.timesofindia.com/Repository/ml.asp?Ref=Q0FQLzIwMTIvMDMvMTQjQXIwMDEwNQ==&Mode=HTML&Locale=english-skin-custom (accessed 14 March 2012).
35. Jack Linchuan Qiu, *Working-Class Network Society: Communication Technology and the Information Have-Less in Urban China* (Cambridge, MA: MIT Press, 2009). Phones aren't meant to hurt people; revolvers are. And revolvers are not much use for playing games or music.
36. Tina Rosenberg, 'Ancient Tongues Meet Digital Age', *International Herald Tribune*, 12 December 2011, pp. 15 and 17. Italics are ours.
37. Eija-Liisa Kasesniemi and Pirjo Rautiainen, 'Mobile Culture', in Katz and Aakhus (eds), *Perpetual Contact*, p. 183. See also Castells *et al.*, *Mobile Communication*, pp. 179–84.
38. Yukiko Nishimura, 'Japanese Keitai Novels and Ideologies of Literacy', in Crispin Thurlow and Kristine Mroczek, *Digital Discourse: Language in the New Media* (London: Oxford University Press, 2011), pp. 86–109.
39. Rosenberg, *International Herald Tribune*, 12 December 2011, pp. 15 and 17.
40. Devanagari (Hindi), Roman (English), Bengali, Gujarati, Gurmukhi (Punjabi), Kannada, Malayalam, Oriya, Tamil, Telugu, Perso-Arabic (Urdu).
41. Booklet, http://www.paninikeypad.com/download/Booklet.pd. See also Nokia Research Centre, 'Solving the Great Indian Text Input Puzzle: Touch Screen-based Mobile Text Input Design', paper presented at MobileHCI, Stockholm, 30 August to 2 September 2011, http://goo.gl/3OGRy (accessed on 12 September 2012). Dr Nimmi Rangaswamy of Microsoft, Bengaluru, led us to these sources.
42. Nimmi Rangaswamy, Microsoft, Hyderabad, email to R. Jeffrey, 16 November 2011.
43. Morozov, *Net Delusion*, pp. 175–6.
44. Marvin, *When Old Technologies Were New*, p. 30.
45. Interview, Bala Parthasarathy, Authentication and Applications, Aadhaar, with R. Jeffrey, Bengalaru, 19 November 2012.
46. *Times of India*, 8 August 2012, http://timesofindia.indiatimes.com/india/Every-poor-family-may-get-a-mobile/articleshow/15395670.cms (accessed 8 August 2012). The cost was estimated at Rs 7,000 crores or about US $1.2 billion. James Wolfensohn, former president of the World Bank, floated similar ideas—a phone for every woman—two months earlier. *New York Times*, 15 June 2012, http://www.nytimes.com/2012/06/16/opinion/a-cellphone-for-every-woman.html (accessed 9 August 2012).
47. http://www.hindu.com/2006/09/16/stories/2006091611920400.htm (accessed 4 April 2012). *Outlook*, 16 July 2012, http://www.outlookindia.com/article.aspx?281538 (accessed 10 August 2012).
48. Madon, *e-Governance for Development*, p. 164.
49. *Documentation of Best Practice. SMS Based Monitoring System* (npp: OneWorld Foundation India, 2010), p. 4, http://indiagovernance.gov.in/bestpractices.php?id=393 (accessed on 12 March 2012).

NOTES

50. Ling and Donner, *Mobile Communication*, p. 152. Madon, *e-Governance*, p. 162.
51. See, for example, http://www.washingtonpost.com/world/asia-pacific/indians-use-cellphones-to-plug-holes-in-governance/2011/10/24/gIQAooAmOM_story.html (accessed 17 April 2012)
52. Dr Vinay Kumar Malaviya, 'Mobile Phone', in Punarnava Magazine, *Dainik Jagran*, vol. 7 (in Hindi, January 2011), p. 17.
53. Nandan Nilekani, *Imagining India: Ideas for the New Century* (London: Allen Lane, 2008), p. 122.
54. For example, Surabhi Mittal, Sanjay Gandhi, Gaurav Tripathi, *Socio-Economic Impact of Mobile Phones on Indian Agriculture*, Working Paper No. 246 (New Delhi: Indian Council for Research on International Economic Relations, 2010). Nilekani, *Imagining India*, pp. 118–19.
55. Fernand Braudel, *Capitalism and Material Life, 1400–1800* (London: Fontana, 1974), p. ix.
56. Braudel, *Capitalism*, p. x.

BIBLIOGRAPHY

Reports, Official Publications and Annual Compendiums

Census of India 2011. Houses. Household Amenities and Assets. Latrine Facility, New Delhi, Registrar General and Census Commissioner, 2011), http://www.censusindia.gov.in/2011census/hlo/Data%20sheet/Latrine.pdf

Census of India 2011. Rural Urban Distribution of Population. Provisional Population Totals, New Delhi, Registrar General and Census Commissioner, 2011, http://censusindia.gov.in/2011-prov-results/paper2/data_files/india/Rural_Urban_2011.pdf

Cellular Operators Association of India, *Annual Report*, various years, http://www.coai.com/annual.php

Comptroller and Auditor-General, *Performance Audit Report on the Issue of Licences and Allocation of 2G Spectrum by the Department of Telecommunications*, New Delhi: CAG, www.thehindu.com/news/article889943.ece?service=mobile (accessed on 20 March 2012).

Department of Posts, *Annual Report*, New Delhi: Department of Posts, India, various years, http://www.indiapost.gov.in/Annual_Reports.aspx

Department of Telecommunications [DoT], *Annual Report*, various years, http://www.dot.gov.in/annualreport/annualreport.htm

India: a Reference Annual, various years. New Delhi: Publications Division.

Information and Communications for Development 2012: Maximizing Mobile (Washington, DC: World Bank, 2012), http://www.worldbank.org/ict/IC4D2012 (accessed 31 July 2012).

Statistical Outline of India, various years. Mumbai: Tata Services.

BIBLIOGRAPHY

Telecom Regulatory Authority of India [TRAI], *Annual Report*, various years, http://www.trai.gov.in/Content/Annual_Reports.aspx

Travancore Administration Report, 1878–9 (1880) Trivandrum: Government Press.

Periodicals and newspapers

Dainik Jagran, Varanasi, daily (Hindi)
Hindu, daily, Chennai
India Today, New Delhi, weekly
International Herald Tribune, daily
My Mobile, New Delhi, monthly (English and Hindi)
Outlook, New Delhi, weekly
Tehelka, New Delhi, weekly
Times of India, daily, Mumbai

Books and Articles

Abraham, R. (2007) 'Mobile Phones and Economic Development: Evidence from the Fishing Industry in India', *Information Technology and International Development*, vol. 4, no. 1, pp. 5–17.

Adhikari, A. and K. Bhatia (2011) 'NREGA Wage Payments: Can We Bank on the Banks', *Economic and Political Weekly*, 2 January, pp. 30–7.

Agar, J. (2003) *Constant Touch: A Global History of the Mobile Phone*, London: Icon Books.

Ahearn, L. (2001) *Invitations to Love: Literacy, Love Letters, and Social Change in Nepal*, Ann Arbor: University of Michigan Press.

Ang, Peng Hwa, S. Tekwani and G. Wang (2012) 'Shutting Down the Mobile Phone and the Downfall of Nepalese Society', *Pacific Affairs*, vol. 85, no. 3, pp. 547–62.

Ashuri, T. (2011) *From the Telegraph to the Computer: a History of Electronic Media*, in Hebrew, Tel-Aviv: Resling Publishing.

Athreya, M. B. (1996) 'India's Telecommunications Policy: a Paradigm Shift', *Telecommunications Policy*, vol. 20, no. 1, pp. 11–22.

Barendregt, B. (2009) 'Mobile Religiosity in Indonesia: Mobilized Islam, Islamized Mobility and the Potential of Islamic Techno Nationalism', in *Living the Information Society in Asia*, ed. Erwin Alampey, Singapore: Institute of Southeast Asian Studies, pp. 73–92.

Basu, S. (1999) *She Comes to Take Her Rights: Indian Women, Property, and Propriety*, Albany: SUNY Press.

Bayly, C. A. (1996) *Empire and Information*, Cambridge: Cambridge University Press.

Bhatt, V. and G. Versaikar (2011) *1857. The Real Story of the Great Uprising*, trans. Mrinal Pande. New Delhi: Harper Perennial.

BIBLIOGRAPHY

Bloch, M. (1961) *Feudal Society*, vol. 1, *The Growth of Ties of Dependence*, London: Routledge and Kegan Paul.

Blum, A. (2012) *Tubes: a Journey to the Center of the Internet*, New York: Ecco/HarperCollins.

Boo, K. (2012) *Behind the Beautiful Forevers*, London: Hamish Hamilton.

Bose, A. (2008) *Behenji: a Political Biography of Mayawati*, New Delhi: Penguin.

Bose, B. (ed.) (2006) *Gender and Censorship*, Delhi: Women Unlimited.

Braudel, F. (1974) *Capitalism and Material Life, 1400–1800*, London: Fontana.

Brinkman, I., M. de Bruijn and H. Bilal (2009) 'The Mobile Phone, "Modernity" and Change in Khartoum, Sudan', in M. de Bruijn *et al.*, *Mobile Phones: The New Talking Drums of Everyday Africa*, Bamenda, Leiden: Langaa RPCIG, African Studies Centre, pp. 69–91.

Bureau of Labour Statistics (2012) 'Automotive Industry: Employment, Earnings, and Hours', February 2012, http://www.bls.gov/iag/tgs/iagauto.htm (accessed 13 April 2012).

Casson, H. N. (2006 [1910]) *The History of the Telephone*, New York: Cosimo Classics.

Castells, M. (2008) 'Afterword', in James E. Katz (ed.), *Handbook of Mobile Communication Studies*, Cambridge, MA: MIT Press, pp 447–51.

Castells, M. (2010 [1996]) *The Rise of the Network Society*, 2nd edition, Chichester: Wiley-Blackwell.

Castells, M., M. Fernandez-Ardevol, J. L. Qiu and A. Sey (2004) *The Mobile Communication Society: A Cross-Cultural Analysis of Available Evidence on the Social Uses of Wireless Communication Technology*, Los Angeles: University of Southern California. http://hack.tion.free.fr/textes/MobileCommunicationSociety.pdf (accessed 5 April 2012).

—— (2007) *Mobile Communication and Society. A Global Perspective*, Cambridge, MA: MIT Press.

'Cell Phone Radiation Warning Ordinance Reinstated in San Francisco' (2011) Filutowski Law Firm, 21 July, http://www.filutowskilaw.com/2011/07/cell-phone-warning-ordinance-reinstated-in-san-francisco/.

Chakravartty, P. (2004) 'Telecom, National Development and the Indian State: a Postcolonial Critique', *Media, Culture and Society*, vol. 26, no. 2, pp. 227–49.

Chandra, K. (2004) *Why Ethnic Parties Succeed: Patronage and Ethnic Head Counts in India*, New York: Cambridge University Press.

Chatterjee, R. (2000) 'Representations of Gender in Folk Paintings of Bengal', *Social Scientist*, vol. 28, nos. 3/4, pp. 7–21.

Choudhury, S. (2010) *Braking News*, Gurgaon: Hachette India.

BIBLIOGRAPHY

Clarke, S. H. (2007) *Trust and Power*, New York: Cambridge University Press.

Cohn, B. S. (1996) 'The Command of Language and the Language of Command', in Bernard S. Cohn, *Colonialism and its Forms of Knowledge*, Princeton: Princeton University Press, pp. 16–56.

Dalgado, M. S. R. (1988) *Portuguese Vocables in Asiatic Languages*, Delhi: Asian Education Services.

Das, G. (2002) *India Unbound*, New York: Alfred A. Knopf.

Dash, M. and A. Mehta (2011) 'Understanding Mobile Phone Radiation and Its Effects', *Economic and Political Weekly*, 23 April, pp. 22–5.

Deloche, J. (1993) *Transport and Communications in India Prior to Steam Locomotion*, vol. 1, *Land Transport*, trans. James Walker, New Delhi: Oxford University Press.

Derné, S. (1994) 'Hindu Man Talk About Controlling Women: Cultural Ideas as the Tool of the Powerful', *Sociological Perspectives*, vol. 37, no. 2, pp. 203–27.

Desai, A. V. (2006) *India's Telecommunications Industry*, New Delhi: Sage.

Dickey, S. (2000) 'Permeable Homes: Domestic Service, Household Space, and the Vulnerability of Class Boundaries in Urban India', *American Ethnologist*, vol. 27, no. 2, pp. 462–89.

Diga, K. (2008) 'Mobile Cell Phones and Poverty Reduction: Technology Spending Patterns and Poverty Level Change among Households in Uganda', written April 2008, Workshop on the Role of Mobile Technologies, Sao Paolo, Brazil, 2–3 June 2008, http://www.w3.org/2008/02/MS4D_WS/papers/position_paper-diga-2008pdf.pdf (accessed 13 January 2012).

Documentation of Best Practice. SMS Based Monitoring System (2010) npp: OneWorld Foundation India, http://indiagovernance.gov.in/best-practices.php?id=393.

Dokeniya, A. (1999) 'Re-forming the State: Telecom Liberalization in India', *Telecommunications Policy*, vol. 23, pp. 105–28.

Donald, S. H., T. D. Anderson and D. Spry (eds) (2010) *Youth, Society and Mobile Media in Asia*, London: Routledge.

Donner, J. (2007) 'Customer Acquisition among Small and Informal Businesses in Urban India: Comparing Face to Face and Mediated Channels', *Electronic Journal on Information Systems in Developing Countries*, vol. 32, pp. 1–16.

Donner, J., and M. X. Escobari (2010) 'A Review of Evidence on Mobile Use by Micro and Small Enterprises in Developing Countries', *Journal of International Development*, vol. 22, pp. 641–58.

Donner, J., N. Rangaswamy, M. Steenson and C. Weil (2008) '"Express Yourself" and "Stay Together": the middle-class Indian family', in E.

J. Katz (ed.), *Handbook of Mobile Communication Studies*, Cambridge, MA: MIT Press, pp. 325–37.

Doron, A. (2008) *Caste, Occupation and Politics on the Ganges: Passages of Resistance*, Farnham: Ashgate.

—— (2010) 'India's Mobile Revolution: a View from Below', *Inside Story*, 10 February, http://inside.org.au/india-mobile-revolution/ (accessed 19 August 2012).

—— (2012) 'Consumption, Technology and Adaptation: Care and Repair Economies of Mobile Phones in North India', *Pacific Affairs*, vol. 85, no. 3, pp. 563–86.

—— (2012) 'Mobile Persons: Cell Phones, Gender and the Self in North India', *The Asia Pacific Journal of Anthropology*, vol. 13 no. 5, pp. 414–433.

Dossani, R. (2008) *India Arriving*, New York: American Management Association.

Dossani, R. (ed.) (2012) *Telecommunications Reform in India*, London: Quorum Books.

Dube, L. (1988) 'On the Construction of Gender: Hindu Girls in Patrilineal India', *Economic and Political Weekly*, 4 June, p. 11–19.

Duggal, P. (2011) *Mobile Law*, New Delhi: Universal Law Publishing Co.

Edgerton, D. (2007) *The Shock of the Old. Technology and Global History since 1900*, New York: Oxford University Press.

'Electromagnetic Fields and Public Health. Base Stations and Wireless Technologies' (2006) World Health Organization, Fact Sheet, No. 304 (May). www.who.int/mediacentre/factsheets/fs 304/en/print.html

Eriksen, T. K. (2007) *Globalization: The Key Concepts*, New York: Berg.

Fernandes, L. (2006) *India's New Middle Class*, Minneapolis: University of Minnesota Press.

Ferus-Comelo, A. and P. Poyhonen (2011) *Phony Equality: Labour Standards of Mobile Phone Manufacturers in India*, Helsinki: Finnwatch, Cividep and SOMO. http://cividep.org/wp-content/uploads/Phony_Equality.pdf (accessed 6 February 2012).

Fischer, C. S. (1992) *America Calling. A Social History of the Telephone to 1940*, Berkeley: University of California Press.

Fischer, J. (2008) *Shopping among the Malays in Modern Malaysia*, Copenhagen: Nordic Institute of Asian Studies.

Fisher, M. H. (1993) 'The Office of Akhbar Nawis: the Transition from Mughal to British Forms', *Modern Asian Studies*, vol. 27, no. 1, pp. 45–82.

Fortunati, L., A. M. Manganelli, P. Law and S. Yang (2008) 'Beijing Calling … Mobile Communication in Contemporary China', *Knowledge, Technology and Policy*, vol. 21, no. 1, pp. 19–27.

BIBLIOGRAPHY

Forty Years of Telecommunications in Independent India (1987) New Delhi: Department of Telecommunications.

Foucault, M. (1997) *Ethics: Subjectivity and Truth*, ed. Paul Rainbow, trans. Robert Hurley and others, New York: New Press.

Gandhi, M. K. *Collected Works of Mahatma Gandhi*, vol. 81, Letter No. 646, 20 December 1941, p. 388.

Gandhi, S., S. Mittal and G. Tripathi (2009) 'The Impact of Mobiles on Agricultural Productivity', in *India: the Impact of Mobile Phones*, Policy Papers Series No. 9, New Delhi: Vodafone Group, pp. 21–33.

Gill, K. (2010) *Of Poverty and Plastic: Scavenging and Scrap Trading Entrepreneurs in India's Urban Informal Economy*, New Delhi: Oxford University Press.

Giridharadas, A. (2011) *India Calling*, Melbourne: Black Inc.

Gleick, J. (2011) *The Information: a History, a Theory, a Flood*, London: Fourth Estate.

Goggin, G. (2006) *Cell Phone Culture*, London: Routledge.

—— (2011) *Global Mobile Media*, London: Routledge.

Gupta, A. (1992) 'The Reincarnation of Souls and the Rebirth of Commodities: Representations of Time in "East" and "West"', *Cultural Critique*, vol. 22, pp. 187–211.

Gurumurthy, A. and N. Menon (2009) 'Violence Against Women via Cyberspace', *Economic and Political Weekly*, 3 October, pp. 19–21.

Hardtmann, E-M. (2009) *The Dalit Movement in India*, New Delhi: Oxford University Press.

Henry, E. O. (1988) 'Social Structure and Music: Correlating Musical Genres and Social Categories in Bhojpuri-Speaking India', *International Review of the Aesthetics and Sociology of Music*, vol. 19, no. 2, pp. 217–27.

Herman, P. (2010) 'Seeing the Divine through Windows: Online Darshan and Virtual Religious Experience', *Heidelberg Journal of Religions on the Internet*, vol. 4, no. 1, pp. 151–178.

Herzfeld, M. (2003) *The Body Impolitic: Artisans and Artifice in the Global Hierarchy of Value*, Chicago: University of Chicago Press.

Hoek, L. (2010) 'Cut-Pieces as Stag Film: Bangladeshi Pornography in Action Cinema', *Third Text*, vol. 24, no. 1, pp. 135–148.

Horst, H. and D. Miller (2005) 'From Kinship to Link-up: Cell Phones and Social Networking in Jamaica', *Current Anthropology*, vol. 46, no. 5, pp. 755–78.

Horst, H. and D. Miller (2006) *The Cell Phone: an Anthropology of Communication*, Oxford: Berg.

Howard, P. N. (2011) *Castells and the Media*, Cambridge: Polity.

Hughes, N. and S. Lonie (2007) 'M-PESA: Mobile Money for the

BIBLIOGRAPHY

"Unbanked": Turning Cellphones into 24-Hour Tellers in Kenya', *Innovations* 2, no. 1–2, pp. 63–81.

Ito, M., D. Okabe and M. Matsuda (eds) (2005) *Personal, Portable, Pedestrian: Mobile Phones in Japanese Life*, Cambridge, MA: MIT Press.

[Iyer, L.] *Documentation on Empowering ASHAs—Mobile Money Transfer, Distt. Shiekpura* [sic]*, BIHAR* (New Delhi: Norway India Partnership Initiative Secretariat, 2010)

Jaffrelot, C. (2003) *India's Silent Revolution: The Rise of the Lower Castes*, London: Hurst.

Jeffrey, Craig, Patricia Jeffery and Roger Jeffery (2008) *Degrees without Freedom? Education, Masculinities, and Unemployment in North India*, Stanford: Stanford University Press.

Jeffrey, Robin, Ronojoy Sen, Pratima Singh (eds) (2012) *More than Maoism*, New Delhi: Manohar for the Institute of South Asian Studies, Singapore.

Jeffrey, Robin (1986) *What's Happening to India?*, London: Macmillan.

——— (2001) '[Not] Being There: Dalits and India's Newspapers', *South Asia*, vol. 24, no. 2, pp. 225–38.

——— (2010) 'The Mahatma Didn't Like the Movies and Why it Matters: Indian Broadcasting Policy, 1920s–1990', in Robin Jeffrey, *Media and Modernity*, New Delhi: Permanent Black, pp. 233–56.

Jeffrey, Robin and Assa Doron (2012) 'Mobile-izing: Democracy, Organization and India's First "Mass Mobile Phone" Elections', *Journal of Asian Studies*, vol. 71, no. 1, pp. 63–80.

——— (2011) 'Celling India: Exploring a Society's Embrace of the Mobile Phone', *South Asian History and Culture*, vol. 2, no. 3, pp. 397–416.

——— (2012) 'The Mobile Phone in India and Nepal: Political Economy, Politics and Society', *Pacific Affairs*, vol. 85, no. 3, pp. 469–82.

Jensen, R. (2007) 'The Digital Provide: Information (Technology), Market Performance and Welfare in the South Indian Fisheries Sector', *Quarterly Journal of Economics*, vol. 122, no. 3, pp. 879–924.

Joshi, B. R. (1981) 'Scheduled Caste Voters: New Data, New Questions', *Economic and Political Weekly*, 15 August, pp. 1357–62.

——— (1987) 'Recent Developments in Inter-Regional Mobilization of Dalit Protest in India', *South Asia Bulletin*, vol. 7, pp. 86–96.

Jung, Y., D. Joshi, V. Narayanan-Saroja and D. P. Desai (2011) Nokia Research Centre, 'Solving the Great Indian Text Input Puzzle: Touch Screen-based Mobile Text Input Design', paper presented at Mobile-HCI, Stockholm, 30 August to 2 September 2011, http://goo.gl/3OGRy (accessed on 12 September 2012).

Kamdar, M. (2007) *Planet India*, New York: Scribner.

BIBLIOGRAPHY

Kasesniemi, E-L. and P. Rautianinen (2002) 'Mobile Culture of Children and Teenagers in Finland', in Katz and Aakhus (eds), *Perpetual Contact*, Cambridge: Cambridge University Press, pp. 170–92.

Kathuria, R., M. Uppal and Mamta [sic] (2009) 'An Econometric Analysis of the Impact of the Mobile', in *India: the Impact of Mobile Phones*, New Delhi: Vodafone, pp. 5–20.

Katz, J. E. (ed.) (2008) *Handbook of Mobile Communication Studies*, Cambridge, MA: MIT Press.

Katz, J. E. and M. Aakhus (eds) (2002) *Perpetual Contact. Mobile Communication, Private Talk, Public Performance,* Cambridge: Cambridge University Press.

Kaul, C. (2004) *Reporting the Raj*, Manchester: Manchester University Press.

Kesavan, B. S. (1985) *History of Printing and Publishing in India*, vol. 1, *South Indian Origins of Printing and Its Efflorescence in Bengal*, New Delhi: National Book Trust of India.

Keskar, D. K. (2009) 'Reliance Infocomm', in A. Mukherjee (ed.), *The Icfai Press on Mobile Service Providers: Perspectives and Practices*, Hyderabad: Icfai University Press, pp. 199–215.

Korom, F. (1998) 'On the ethics and aesthetics of recycling in India', in Lance Nelson (ed.), *Purifying the Earthly Body of God: Religion and Ecology in Hindu India*, New York: New York University Press, pp. 197–223.

Kumar, A., A. Tewari, G. Shroff, D. Chittamuru, M. Kam, and J. Canny (2010) 'An Exploratory Study of Unsupervised Mobile Learning in Rural India', paper presented at CHI 2010 conference, Atlanta, 10–15 April 2010, http://www.cs.cmu.edu/~anujk1/CHI2010.pdf (accessed 28 February 2012).

Kumar, M. and R. K. Kakani (2012) *The Telecommunications Revolution: Mobile Value Added Services in India*, New Delhi: Social Science Press.

Kumar, P. and R. Golait (2009) 'Bank Penetration and SHG-Bank Linkage Programme: a Critique', *Reserve Bank of India Occasional Papers*, vol. 29, no. 3, pp. 119–38.

Kumar, V. (2006) *India's Roaring Revolution: Dalit Assertion and New Horizons*, Delhi: Gagandeep Publications.

Lahiri Choudhury, D. K. (2010) *Telegraphic Imperialism*, Houndmills: Palgrave Macmillan.

Lamb, S. (1997) 'The Making and Unmaking of Persons: Notes on Aging and Gender in North India', *Ethos*, vol. 25, no. 3, pp. 279–302.

Lang, J. (2008) *Legends of India: Tales of Life in Hindostan*, ed. by Victor Crittenden, Canberra: Mulini Press.

Lazarsfeld, P. F., B. Berelson, and H. Gaudet (1944) *The People's Choice:*

BIBLIOGRAPHY

How the Voter Makes Up His [sic] *Mind in a Presidential Campaign*, New York: Columbia University Press.

Leacock, S. (1910) 'My Financial Career', in *Literary Lapses*, Montreal: Gazette Printing Company, pp. 5–9.

Ling, R. (2004) *The Mobile Connection: the Cell Phone's Impact on Society*, San Francisco: Elsevier.

Ling, R. and J. Donner (2009) *Mobile Communication*, Cambridge: Polity.

Ling, R. and S. W. Campbell (2011) *Mobile Communication: Bringing Us Together and Tearing Us Apart*, New Brunswick, NJ: Transaction Publishers.

Logan, W. (1887) *Malabar*, vol. 1, Madras: Government Press, reprint 1951.

Lu, J. and I. Weber (2007) 'State, Power and Mobile Communication: A Case Study of China', *New Media and Society*, vol. 9, no. 6, pp. 925–44.

MacKinnon, R. (2012) *Consent of the Networked: The Worldwide Struggle for Internet Freedom*, New York: Basic Books.

Madon, S. (2009) *e-Governance for Development: a Focus on Rural India*, Houndmills: Palgrave Macmillan.

Malaviya, V. K. (2011) 'Mobile Phone', in Punarnava Magazine, *Dainik Jagran*, vol. 7 (in Hindi), p. 17.

Mani, S. (2012) 'The Mobile Communications Services Industry in India: Has it led to India Becoming a Manufacturing Hub for Telecommunication Equipment?', *Pacific Affairs*, vol. 85, no 3 (September 2012), pp. 511–30.

Manucci, N. (1907) *Mogul India, 1653–1708*, vol. 2, trans. William Irvine, London: John Murray.

Manuel, P. (1993) *Cassette Culture: Popular Music and Technology in North India*, Chicago: Chicago University Press.

Markovits, C. (2006) 'Merchant Circulation in South Asia', in C. Markovits, J. Pouchepadass and S. Subrahmanyam (eds), *Society and Circulation*, London: Anthem Press, pp. 131–162.

Martin, M. (1991) *'Hello, Central?' Gender, Technology and Culture in the Formation of Telephone Systems*, Quebec: McGill-Queen's University Press.

Martinez-Jerez, F. A. and V. G. Narayanan (2007) 'Strategic Outsourcing at Bharti Airtel Limited', *Harvard Business School, 9–107–003*, 2006 (revised 4 December 2007).

Marvin, C. (1988) *When Old Technologies Were New: Thinking about Electronic Communication in the Late Nineteenth Century*, Oxford: Oxford University Press.

Mazumdar, R. (2011) 'Terrorism, Conspiracy, and Surveillance in Bombay's Urban Cinema', *Social Research*, vol. 78, no. 1, pp. 143–172.

BIBLIOGRAPHY

Mazzarella, W. (2003) *Shoveling Smoke: Advertising and Globalization in Contemporary India*, Durham, NC: Duke University Press.

McDonald, H. (2010) *Mahabharat in Polyester*, Sydney: University of New South Wales Press.

McEwen, R. M. and Melissa E. Fritz (2011) 'EMF Social Policy and Youth Mobile Phone Practices in Canada', in J. E. Katz (ed.), *Mobile Communication*, New Brunswick, NJ: Transaction Publishers, pp. 133–56.

Mehta, Nalin. (2008) *India on Television*, New Delhi: HarperCollins.

Mehta, Vinod (2011) *Lucknow Boy*, New Delhi: Penguin Viking.

Melody, W. H. (2000) 'Spectrum Management for Information Societies', in John Ure (ed.), *Telecommunications Development in Asia*, Hong Kong: Hong Kong University Press, pp. 57–84.

Metcalf, B. D. (1982) *Islamic Revival in British India: Deoband, 1860–1900*, Princeton: Princeton University Press.

Miller, Daniel (2011) *Tales from Facebook*, Cambridge: Polity.

Miller, Daniel and D. Slater (2005) 'Comparative Ethnography of New Media', in James Curran and Michael Gurevitch (eds), *Mass Media and Society*, 4th edition, London: Hodder Arnold, pp. 303–319.

Miller, Eric J. (1954) 'Caste and Territory in Malabar', *American Anthropologist*, vol. 56, no. 3, pp. 410–420.

Mishra, P. (2006 [1995]) *Butter Chicken in Ludhiana*, London: Picador.

Mittal, S. B. (2006) *India's New Entrepreneurial Classes: the High Growth and Why It Is Sustainable*, Occasional Paper No. 25, Philadelphia: Centre for the Advanced Study of India, University of Pennsylvania.

Mittal, S., S. Gandhi and G. Tripathi (2010) *Socio-Economic Impact of Mobile Phones on Indian Agriculture*, Working Paper No. 246, New Delhi: Indian Council for Research on International Economic Relations.

Mittal, S., S. Gandhi and G. Tripathi (2009) 'The Impact of Mobile Phones on Agricultural Productivity", in Kathuria *et al.*, *India: the Impact of Mobile Phones*, pp. 21–33.

Morozov, E. (2011) *The Net Delusion*, London: Allen Lane.

Mukherji, R. (2009) 'Interests, Wireless Technology and Institutional Change: From Government Monopoly to Regulated Competition in Indian Telecommunications', *Journal of Asian Studies*, vol. 66, no. 2, pp. 491–517.

Muthuswamy and Brinda [no initials] (eds) (1993 [1989]), *Swamy's Treatise on Telephone Rules with Act* [sic], *Rules, Orders, Digest, Guidelines and Case-Law*, Madras: Swamy Publishers.

Nair, C. N. N. (2002) *The Story of Videsh Sanchar: Development of India's External Telecommunications*, Mumbai: Videsh Sanchar Nigam Ltd.

Narayan, B. (2011) *The Making of the Dalit Public in North India*, New Delhi: Oxford University Press.

BIBLIOGRAPHY

Narayan, R. K. (1993) *The Grandmother's Tale*, London: Heinemann.

Nilekani, N. (2008) *Imagining India: Ideas for the New Century*, London: Allen Lane.

Nishant Shah, N. (2007) 'Subject to Technology: Internet Pornography, Cyber-Terrorism and the Indian State', *Inter Asia Cultural Studies Journal*, vol. 8, no 3, pp. 349–66.

Nishimura, Y. (2011) 'Japanese Keitai Novels and Ideologies of Literacy', in C. Thurlow and K. Mroczek, *Digital Discourse: Language in the New Media*, London: Oxford University Press, pp. 86–109.

Norris, L. (2010) *Recycling Indian Clothing: Global Contexts of Reuse and Value*, Bloomington: Indiana University Press.

Nyamnjoh, F. B. (2009) 'Married but Available', excerpt in M. de Bruijn, F. Nyamnjoh and I. Brinkman (eds), *Mobile Phones: the New Talking Drums of Everyday Africa*, Leiden: African Studies Centre, pp. 1–10.

O'Neill, P. D. (2003) 'The "Poor Man's Mobile Telephone": Access versus Possession to Control the Information Gap in India', *Contemporary South Asia*, vol. 12, no. 1, pp. 85–102.

Olivelle, P. (ed. and trans.) (1999) *Dharmasutra. The Law Codes of Ancient India*, Oxford: Oxford University Press.

Omari-Hijazi, H. and R. Ribak (2008) 'Playing with Fire: On the Domestication of the Mobile Phone among Palestinian Teenage Girls in Israel', *Information, Communication & Society*, vol. 11, no. 2, pp. 149–66.

Pai, S. (2002) *Dalit Assertion and the Unfinished Democratic Revolution: the Bahujan Samaj Party in Uttar Pradesh*, New Delhi: Sage.

Perschbacher, G. (1996) *Wheels in Motion: the American Automobile Industry's First Century*, Iola, WI: Krause Publications.

Pillai, T. S. (1964) *Chemmeen*, Bombay: Jaico.

Pinney, C. (1999) 'On Living in the Kal(i)yug: Notes from Nagda, Madhya Pradesh', *Contributions to Indian Sociology*, vol. 33, no. 1 & 2, pp. 77–106.

Pitroda, S. (1993) 'Development, Democracy and the Village Telephone', *Harvard Business Review*, vol. 71, no. 6, pp. 66–79. http://groups.google.com/group/soc.culture.indian/browse_thread/thread/c0e82 3d72686cd64/6bb8fa00c1d2790f (accessed 22 October 2010).

Pitroda, S. and Mehul Desai (2010) *The March of Mobile Money: the Future of Lifestyle Management*, New Delhi: Collins Business.

Plouffe, D. (2010) *The Audacity to Win*, New York: Viking.

Portus, L. M. 'How the Urban Poor Acquire and Give Meaning to the Mobile Phone', in J. E. Katz (ed.), *Handbook of Mobile Communication Studies*, Cambridge, MA: MIT Press, p. 105–18.

Powell, J. (1995) *The Survival of the Fitter. Lives of Some African Engineers*, Rugby: Intermediate Technology Publications.

BIBLIOGRAPHY

Prahalad, C. K. (2005) *The Fortune at the Bottom of the Pyramid: Eradicating Poverty through Profits*, Upper Saddle River, NJ: Wharton School Publishing.

Purkayastha, P. (1996) 'Induction of Private Sector in Basic Telecom Services', *Economic and Political Weekly*, 17 February, p. 417.

Qiu, J. L. (2009) *Working-Class Network Society: Communication Technology and the Information Have-Less in Urban China*, Cambridge, MA: MIT Press.

Radhika, A. Neela and A. Mukund (2003) 'Airtel Magic—Selling a Prepaid Cellphone Service', ICFAI University Press, 2003, http//www.scribd.com/doc/54318936/Airtel-Magic-Selling-a-Pre-Paid-Cellphone-Service (accessed on 21 September 2012).

Rafael, V. L. (2003) 'The Cell Phone and the Crowd: Messianic Politics in the Contemporary Philippines', *Public Culture*, vol. 15, no. 3, pp. 399–425.

Raha, D. and S. Cohn-Sfetcu (2009) 'Turning the Cellphone into an Antipoverty Vaccine', *Journal of Communications*, vol. 4, no. 3, pp. 203–10.

Raheja, G. G. and A. G. Gold (1994) *Listen to the Heron's Words: Reimagining Gender and Kinship in North India*, Berkeley: University of California Press.

Rainie, L. and B. Wellman (2012) *The New Social Operating System*, Cambridge, MA: MIT Press.

Raja, I. (2009) 'Rethinking Relationality in the Context of Adult Mother-Daughter Caregiving in Indian Fiction', *Journal of Aging, Humanities and the Arts*, vol. 3, pp. 25–37.

Rajagopal, A. (2001) 'The Violence of Commodity Aesthetics: Hawkers, Demolition Raids, and a New Regime of Consumption', *Social Text*, vol. 19, no. 3, pp. 91–113.

Rangarajan, M. (2008) 'Why Mayawati Matters', *Seminar*, no. 581 (January 2008).

Rangaswamy, N. and S. Nair (2010) 'The Mobile Phone Store Ecology in a Mumbai Slum Community: Hybrid Networks for Enterprise', *Information Technologies & International Development*, vol. 6, no. 3, pp. 51–65.

Rao, V. N. and S. Subrahmanyam (2003) 'Circulation, Piety and Innovation: Recounting Travels in Early Nineteenth-Century South India', *Society and Circulation*, New Delhi: Permanent Black, pp. 306–56.

Ravindran, G. (2009) 'Moral Panic and Mobile Phones: The Cultural Politics of New Media Modernity in India', in Erwin Alampay (ed.), *Living the Information Society in Asia*, Singapore: Institute of Southeast Asian Studies.

Rawlley, V. (2010) 'Miss Use: A Survey of Raunchy Bhojpuri Music Album', *TasveerGhar: A Digital Archive of South Asian Popular Visual*

BIBLIOGRAPHY

Culture, http://tasveergharindia.net/cmsdesk/essay/66/index.html (accessed 10 May 2010).

Rheingold, H. (2002) *Smart Mobs: the Next Social Revolution*, New York: Basic Books.

Riello, G. and P. McNeill (2006) 'A Long Walk: Shoes, People and Places', in Giogio Riello and Peter McNeill (eds), *Shoes: a History from Sandals to Sneakers*, New York: Berg, pp. 2–29.

Rosenberg, T. (2011) 'Ancient Tongues Meet Digital Age', *International Herald Tribune*, 12 December, pp. 15 and 17.

Saha, A. (2012) 'Cellphones as a Tool for Democracy: The Example of CGNet Swara', *Economic and Political Weekly*, 14 April, pp. 23–6.

Sarkar, S. (1992) '"Kaliyuga", "Chakri" and "Bhakti": Ramakrishna and His Times', *Economic and Political Weekly*, 18 July, pp. 1543–66.

Sarkar, T. (2001) *Hindu Wife, Hindu Nation: Community, Religion, and Cultural Nationalism*, London: Hurst.

Saxena, D. (2009) *Connecting India: Indian Telecom Story*, New Delhi: Konark.

Schimmel, A. (2004) *The Empire of the Great Mughals*, London: Reaktion Books.

Shamir, O. and G. Ben-Porat, (2007) 'Boycotting for Sabbath: Religious Consumerism as a Political Strategy', *Contemporary Politics*, vol. 13, no. 1, pp. 75–92.

Shankar, K. (2009) 'Sexting: A New Form of Victimless Crime?', *International Journal of Cyber Criminology*, vol. 3. no. 1, pp. 21–5.

Sharma, D. C. (2009) *The Long Revolution. The Birth and Growth of India's IT Industry*, New Delhi: HarperCollins.

Sharma, D. and A. K. Ojha (2011) 'Evolution of the Indian Mobile Telecommunications Industry: Looking through the C-Evolutionary Lens', paper presented at the 'Celling South Asia' workshop, Institute of South Asian Studies, Singapore, 17 and 18 February 2011.

Sharma, K. *et al.* (2009) 'BSNL: Ringing Change in Rural India', in Arindam Mukherji (ed.), *The Icfai University Press on Mobile Service Providers*, Hyderabad: Icfai University Press, pp. 216–41.

Shirky, C. (2008) *Here Comes Everybody: How Change Happens When People Come Together*, London: Penguin.

Shourie, A. (2007 [2004]) *Governance and the Sclerosis That Has Set In*, New Delhi: Rupa.

Silverstone, R. and E. Hirsch (eds) (1992) *Consuming Technologies: Media and Information in Domestic Spaces*, London: Routledge.

Singh Grewal, D. (2008) *Network Power: The Social Dynamics of Globalization*, New Haven: Yale University Press.

Singh, B. (2008) 'Aadamkhor Haseena (The Man-Eating Beauty) and the

Anthropology of a Moment', *Contributions to Indian Sociology*, vol. 42, no. 2, pp. 249–79.

Singh, M. K. and V. Ansari (2011) 'Dharmkagranth' (the book of dharma), in *My Mobile Magazine* (in Hindi) (September), pp. 15–17.

Singh, U. K. (2007) *The State, Democracy and Anti-Terror Laws in India*, New Delhi: Sage.

Smith, J. H. (2011) 'Tantalus in the Digital Age: Coltan Ore, Temporal Dispossession, and "Movement" in the Eastern Democratic Republic of Congo', *American Ethnologist*, vol. 38, no. 1, pp. 17–35.

Smith, R. T. (1996) *The Matrifocal Family: Power, Pluralism and Politics*, New York and London: Routledge.

Southwood, R. (2008) *Less Walk, More Talk: How Celtel and the Mobile Phone Changed Africa*, Chichester: Wiley for Celtel.

Sreekumar, T. T. (2011) 'Mobile Phones and the Cultural Ecology of Fishing', *Information Society*, vol. 27, no. 3, pp. 172–80.

Sridhar, V. (2012) *The Telecom Revolution in India: Technology, Regulation and Policy*, New Delhi: Oxford University Press.

Srivastava, S. (2007) *Passionate Modernity: Sexuality, Class and Consumption in India*, London: Routledge.

Subramanian, D. (2010) *Telecommunications Industry in India*, New Delhi: Social Science Press.

Sundaram, R. (2010) *Pirate Modernity: Delhi's Media Urbanism*, New Delhi: Routledge.

Swamy, S. (2012) *2G Spectrum Scam*, New Delhi: Har-Anand.

Tandon, P. (1981) *Return to Punjab*, Berkeley: University of California Press.

Tenhunen, S. (2008) 'Mobile Technology in the Village: ICTs, Culture and Social Logistics in India', *Journal of the Royal Anthropological Society* (New Series), vol. 14, pp. 515–34.

Thakurta, Paranjoy Guha (2012) *The Great Indian Telecom Robbery. A documentary film on why the misallocation and undervaluation of 2G spectrum is the biggest scandal in independent India (and perhaps the world)* DVD, produced and directed by Paranjoy Guha Thakurta.

Thiruvengadam, A. K. and Piyush Joshi (2012) 'Judiciaries as Crucial Actors in Southern Regulatory Systems: A Case Study of Indian Telecom Regulation', *Regulation and Governance* (2012), http://onlinelibrary.wiley.com/doi/10.1111/j.1748-5991.2012.01143.x/abstract (accessed 20 July 2012).

Thomas, P. N. (2012) *Digital India: Understanding Information, Communication and Social Change*, New Delhi: Sage.

Thompson, E. C. (2009) 'Mobile phones, communities and social networks among foreign workers in Singapore', *Global Networks*, vol. 9, no. 3, pp. 359–80.

BIBLIOGRAPHY

Thompson, E. P. (1967) 'Time, Work-Discipline, and Industrial Capitalism', *Past and Present*, vol. 38, pp. 56–97.

Tiengtrakul, C. (2006) 'Home: Banarasi Women and Perceptions of the Domestic Domain', in Lina Fruzzetti and Sirpa Tenhunen (eds), *Culture, Power, and Agency: Gender in Indian Ethnography*, Kolkata: Stree, pp. 22–51.

Trehan, M. (2009) *Tehelka as Metaphor*, New Delhi: Roli.

Tripathy, R. (2012) 'Music Mania in Small-town Bihar: Emergence of Vernacular Identities', *Economic and Political Weekly*, 2 June, pp. 58–66.

Turner, E. S. (1953) *The Shocking History of Advertising*, New York: Ballantine Books.

Uberoi, P. (2006) *Freedom and Destiny: Gender, Family, and Popular Culture in India*, New Delhi: Oxford University Press.

Umble, D. Z. (1992) 'The Amish and the Telephone: Resistance and Reconstruction', in Roy Silverstone and Eric Hirsch (eds), *Consuming Technologies: Media and Information in Domestic Spaces*, London: Routledge, pp. 171–81.

Uppal, M. and R. Kathuria (2009) 'The Impact of Mobiles in the SME Sector', in *India: the Impact of Mobile Phones*, New Delhi: Vodafone, pp. 54–60.

Uppal, M. with S. K. N. Nair and C. S. Rao (2006) *India's Telecom Reform: a Chronological Account*, New Delhi: National Council for Applied Economic Research.

Ureta, S. (2008) 'Mobilizing Poverty?: Mobile Phone Use and Everyday Spatial Mobility among Low-income Families in Santiago, Chile', *The Information Society*, vol. 24, no. 2, pp. 83–92.

Veeraraghavan, R., N. Yasodhar and K. Toyama (2009) 'Warana Unwired: Replacing PCs with Mobile Phones in a Rural Sugarcane Cooperative', *Information Technologies and International Development*, vol. 5, no. 1, pp. 81–95.

Wallis, C. (2010) 'The Traditional Meets the Technological: Mobile Navigations of Desire and Intimacy' in S. H. Donald, T. D. Anderson and D. Spry (eds), *Youth, Society, and Mobile Media in Asia*, London: Routledge, pp. 57–69.

Walsh, J. (2004) *Domesticity in Colonial India: What Women Learned When Men Gave Them Advice*, Oxford: Rowman and Littlefield.

Walton, M. and J. Donner (2009) 'Red-Write-Erase: Mobile-Mediated Publics in South Africa's 2009 Elections', paper presented at the International Conference on Mobile Communication and Social Policy, Rutgers University, New Brunswick, NJ, 9–11 October 2009, http://research.microsoft.com/en-us/people/jdonner (accessed July 2010).

Wellman, B., (2001) 'Physical Place and Cyberplace: the Rise of Personalized Networking', *International Journal of Urban and Regional Research*, vol. 25, no. 2, pp. 227–52.

Wellman, B. (n.d.; c. 2002) 'Little Boxes, Glocalization, and Networked Individualism', http://homes.chass.utoronto.ca/~wellman/publications/littleboxes/littlebox.PDF.

White, R. (2011) *Railroaded: The Transcontinentals and the Making of Modern America*, New York: W. W. Norton.

INDEX

1857, 2, 21, 25
2G, 57, 60, 89, 203
'2G scandal', 60–1
3G, 52
4G, 209, 218

'A Walled and Surveilled World', 204
A Wednesday, 201
Aadhaar, 221
aam aadmi, 28, 204
Aamir, 201
Accredited Social Health Activists (ASHA), 130
Adler, Mans, 160
advertising, 26, 30, 62, 66–70, 75, 79, 80, 86, 105, 187
affirmative action, 30
Afghanistan, 37
Africa, 3, 5, 11, 14, 54, 116, 121, 158–9, 167, 216
African National Congress, 158
Aggarwal, Anuradha, 71, 105
Agra, 21
agricultural labourer, wage of a, 141
agriculture, 103, 137, 141

Ahearn, Laura, 179
air pollution, 213
Aircel, 56, 236n
Airtel, 33, 54, 66–7, 72–5, 94, 131, 223; *see also* Bharti Airtel
Ajeet Singh, 101–02
Akbar, Emperor, 20
Akhilesh Yadav, 157
Alcatel, 93
Alexander the Great, 20
All India Backward and Minority Communities Employees Federation (BAMCEF), 146–7
All India Post Graduate Medical Entrance examination, 200
All India Radio (AIR), 31, 39
Allahabad, 24, 77, 174, 178
Allahabad University, 77
Alpha Institute, 103
Amazon, 215
Ambani family, 51–2, 56, 70, 74–5
Ambani group, 202
Ambani, Anil, 54
Ambani, Dhirubhai, 39, 54
Ambani, Mukesh, 54, 202
Ambassador car, 39–40
Ambedkar, B. R., 146–7

281

INDEX

America Calling, 89
Amish, 168
Ananda Ranga Pillai, 23
Andhra Pradesh, 161, 204, 222
Anna Karenina, 216
Apollo Hospitals, 56
'Arab spring', 221
Arabic, 219
Arora, Kapil, 68
asbestos, 212
Asian Games, 1982, 29
Assi Ghat, Banaras, 82, 86, 132, 135–6
Athreya Committee, 44
Athreya, M. B., 43
auctions, 36, 45–6, 49, 50–52, 54, 57, 60
Australia, ix-x, xiv, 7, 15, 31, 104, 206, 227n, 233n, 237n
automobile industry, 77, 123; *see also* cars
Awadh, 21
Ayyash, Yahya, 205

Babu, Satish, *see* Satish Babu
Backward Classes, 157
Badrinath, 22
Bahujan Samaj Party (BSP), 8, 147–57, 163, 220
Baiga people, 162
Baltimore, 205
Bambuser, 159–60
BAMCEF, 150
Banaras Hindu University (BHU), 80
Banaras, 9, 65, 79, 81–4, 98–101, 103–04, 107, 123, 132–3, 136, 141, 167, 169, 170–1, 174–5, 178, 180, 187, 192–3, 213 *see also* Varanasi
Bangalore, 21, 37, 92, 196, 221; *see also* Bengalaru

Bangladesh, 37, 54
bank accounts, 116–17, 125; women and, 130; branches, 126; basic, 128; in Kenya, 123
bans on phones, 165, 167, 174, 195, 225n, 256n
Barabanki, 157–8
Barelvi, Sayyid Ahmed, 22
Barendregt, Bart, 13
Basque separatists, 145
BBC, 160
bees, 212
Behram Baug, 100
Beijing, 205
Bell telephone company, 67
Bell, Alexander Graham, 26, 202
Below the Poverty Line (BPL), 222
Benegal, Shyam, 29
Bengal Gazette, 23
Bengal, 22, 190
Bengalaru, 21, 92, 195–6, 221; *see also* Bangalore
bhaaichaara samitis, 152–5
Bharat Electronics Ltd (BEL), 30
Bharat Sanchar Nigam Ltd (BSNL), 50–1, 55–7, 78, 93, 96
Bharatiya Janata Party (BJP), 47, 49, 152, 196
Bharti Airtel, 53–4, 77, 140; *see also* Airtel
Bhojpuri, 82, 106, 187–9, 217, 254n
bicycles, ix, 116, 147, 149, 158, 163, 220
Bihar, 116–17, 123, 126, 128–9, 166, 190, 199, 217, 222, 254n, 258n
Bilaspur, 161
Birla Institute of Technology, 123
Birla, 55–6
Birla, Aditya, 55

282

INDEX

Blackberry, 102, 208
Bloch, Marc, 20, 22
Block Development Officer (BDO), 222
Bluetooth, xxxii, 192–4, 199–200, 217, 219
boatmen at Banaras, 9, 82, 132–5, 169
Bollywood, 104–05, 181, 193, 209, 214
Boston Globe, 9
bottom of the pyramid (BOP), 6, 87, 119, 199
Brahmins, 2, 21–3, 25, 149, 152–4, 163, 189
brandy: needed by printers, 24
Braudel, Fernand, 223–4
bribery, 8, 129, 131, 201, 246n; digital, 223
brides and mobiles, 11, 169–70, 173, 175, 177, 181, 204
Britain, 3–4, 20, 24, 69
broadcast: ability to, 8, 10, 162, 169, 195, 197, 200, 202, 204, 221, 223
Bushmen, 168–9
Butter Chicken in Ludhiana, 53

cable, 35, 42, 51, 57, 138, 246n; needs estimated, 35
California, 65
calls: cost of, x, 12, 46, 67, 72, 141, 151, 210
cameras: in phones, 9, 76, 99, 109, 156, 159, 194–6, 250n
Canada, 35, 168, 191
cancer, 212
car industry, 89–90, 96, 100, 102; *see also* automobile
Caribbean, 11, 121
Cash, Johnny, 14
caste Hindus, 154, 163

caste, 3, 14–15, 19, 22, 30–1, 78, 109, 118, 143, 146–8, 152, 154, 157, 163, 166–7, 174, 176, 181, 188, 210, 214, 220, 231n, 249n, 254n
Castells, Manuel, 9, 134, 145, 209, 215
Catholic Church, 9
CDMA, 51, 54–5, 74
CDs/VCDs/DVDs, 187–8, 191, 193, 195, 217
cell phones, *see* mobiles, phones, teledensity
Cellular Operators Association of India (COAI), 74–5
census, 6, 148, 232n
Central Bureau of Investigation (CBI), 47, 58
Centre for Development of Telematics (C-DOT), 33
CGNet Swara, 160–2, 219
Chandigarh, 131
Charan Singh, 31
Cheeka, 67–9, 216, 238n
Chemmeen, 120
Chennai, 26, 39–40, 43, 55–6, 75, 78, 91, 96
Chhattisgarh, 160, 162, 219
Chicago, University of, 125
Chile, 169
China mobiles, 80, 99, 102, 107, 197, 242n, 249n
China, 3, 6, 13, 30–1, 75, 148, 169, 204, 208, 216, 239n
Chopra, Priyanka, 105
Chorwad, 54
Choudhary, Shubhranshu, 160–2
Christians, 12, 118
Churasia, Sumit, 65, 83–4
Chaze Mobile, 191
ClearTrail, 204
Clinton, Bill, 222

283

INDEX

Coca-Cola, 168
Code Division Multiple Access (CDMA), *see* CDMA
'Code of Morals, A', 204
Cohn, Bernard S., 23
Communications, Department of, 42–5
Communications, Minister of, 5, 48, 57–8, 202–03
Comptroller and Auditor-General (CAG), 58–9
Congress Party, 35, 42, 46–7, 53, 60, 152, 157, 221–2
Constitution of India, 30, 143, 145
consumer goods, 6, 87, 213
consumer practices, 61, 70–2, 77–9, 84–6, 99–100, 108–10, 193, 195, 198, 214; *see also* domestication
consumerism, 76, 110, 187, 191, 210
Consuming Technologies, 168
consumption, 6–7, 11, 104, 108
copper, 74; cost of, 42
crime, 197–202
Crossette, Barbara, 39–40
Cumberland, Duke of, 37
cyber-utopianism, 146
Cyrillic, 219

Daal Mandi, 99–101, 192
dak, 30
Dalits, 147–50, 152–6, 163, 202
Dashashvamedh Ghat, 132
Defence, Minister of, 8
Delhi police, 198, 204
Delhi Public School, 194
Delhi, 43, 67, 72, 95, 100, 167, 198, 214; *see also* New Delhi
Deloche, Jean, 23
Denmark, 13
Desai, Ashok, 40, 46, 49

Dev D, 194
Devanagari, 154
Deve Gowda, H. D., 31
Dharmasutra, 143
Dhirubhai Ambani Entrepreneurs (DAE), 70
dhobis, 223
diabetes, 211
diesel, 94–6, 120–1, 123, 213, 241n
Dinesh Lal, 188
Diwali, 78, 82–3, 85
DMK, *see* Dravida Munnetra Kazhagam
dogs, 67–8; *see also* Cheeka
domestication, 10, 168, 229n; *see also* consumer practices
Donner, Jonathan, 115, 121–3, 158, 200
Dravida Munnetra Kazhagam, 60, 202
Dupleix, 23

East India Company, 23
Economic Weekly, 27
Economist, 119
Egypt, 160
EKO, xxvii, 123–31, 138, 140–1, 246n
Election Commission of India (ECI), 156
electronic waste, 213–15
'emergency' 1975, 24
employment in telecom: estimate of, 61, 89–90, 111, 122–3; women and, 90–4
Engels, Friedrich, 94
English, 102
Escobari, Marcela X., 115, 121–3
Essar, 55–6; *see also* Vodafone, Hutchison
Estrada, President Joseph, 144, 221

INDEX

Europe, 10, 20, 22–3, 27, 37, 52, 92, 110, 208, 216, 219, 239n
e-waste, 213–15

Facebook, 182–3
farmers, 8, 115, 118, 223; mobiles and, 138–41, 247n
Fast Moving Consumer Goods (FMCG), 5, 71–2, 86–7, 90, 124, 131
Federal Communications Commission (FCC), 36
Filipinos, 144
Finland, 14, 219
First Gulf War, 34–5, 42
First World War, 89, 95
Fischer, Claude S., 89, 168
fishermen, 8, 144, 244n; in Kerala, 118–23, 132, 144, 223, 243n
Five Year Plans, 31–2
flash mobs, 221
Fly, 80
footwear, 2; *see also* shoes
Forbes, 55
Ford, Henry, 77, 102, 111
Fortune at the Bottom of the Pyramid, The, 119
Foucault, Michel, 179
Foxconn, 92
France, 4
gaming, xxxii, 217–18, 262n
Gandhi, Indira, 24, 31, 33
Gandhi, M. K., 26–7
Gandhi, Rajiv, 32–5, 39, 40, 42, 54, 56, 90, 137
Ganesh, 189–90
Ganga River, 1, 3–4, 123, 131–2, 225n
garibi hatao, 124
GDP: share of agriculture in, 137
Geekani, Syed Abdul Rehman, 206

gender relations, 11, 14–15, 106, 166–9, 171–2, 177, 182, 190, 210, 217, 231n, 252n, 253n
Genghis Khan, 20
Germany, 4
Ghaffar Market, 100, 103–04, 193
Ghana, 5, 159
ghatwar, 132–3
Giridharadas, Anand, 53, 182
Glasgow, 206
global village, 141
Goa, 68, 116, 244n
Global Positioning Technology, 95
Gods Must Be Crazy, The, 168
Gondi, 160–1, 219
Google Earth, 73
Gorakhpur, 81, 84
gossip, 21, 175, 179, 186–7, 202, 205
Gram Vaani, 219
Grandmother's Tale, The, 21
Grewal, David Singh, 134, 162, 217
grey markets, 75, 86, 98–9, 107–08, 110, 197, 242n
Gross Domestic Product, 119
Global System for Mobile communications (GSM), 54–5, 235n, 238n, 239n
GSM, *see* Global System for Mobile communication
Gujarat, 54, 133, 155
GupShup, 220
Gupta, Ranjeet, 116–17
Guptas, 78

handset, cost of, 141; *see also* phone, mobile
Haneef, Muhamed, 206
Harvard Business Review, 19
Harvard University, 43

285

INDEX

Haryana, 204
Hastings, Warren, 23
Havas, 4
Hazratganj, 97
health, 211–13
Heatwole, Anneryan, 167
heliograph, 204
henna, 116–17
Henry VIII, 165
Hicky, James, 23
hierarchy, social, 3, 137, 152, 174, 210
Himachal Futuristic Communications Ltd (HFCL), 46
Himachal Pradesh, 6, 46
Himalayas, 22
Hind Swaraj, 26
Hindi, 15, 28, 96, 102, 105–7, 150, 199, 220, 223
Hindu cosmology, 190, 214
Hindu values, 196
Hindu, 204
Hindustan Times, 198
Holy Qur'an, 12
Home Ministry, 203
Hong Kong, 55, 67
Horst, Heather, 10, 31, 122, 182
Hutch, 75
Hutchison Whampoa, 55, 67
Hutchison, 69, 71, 75; see also Hutchison Whampoa
Hutchison-Essar group, 67–8; see also Essar
Hyderabad, 217, 222
Hyundai, 40

IBN Live, 207
Ibrahim, Mo, 3
ICICI Bank, 124, 131
Idea, 55, 94
IFFCO Kisan Sanchar Ltd (IKSL), 140–1

illiteracy, 15, 20, 24–5, 29, 125, 127, 134, 139, 150, 154
IMEI number, 102, 117, 197
Income Tax department, 202
India Calling, 53
India Telecom Infra Ltd (ITIL), 95
India Today, 25, 166–7, 196
India's Telecommunications Industry, 40
Indian Post and Telegraphs (IP&T), ix, 28, 29, 148, 150
Indian Telecom Industries (ITI), 92–3
Indonesia, 13, 219
Indore, 204
Indus Towers, 94
industrialization, 20
'information have-less', 218
Infrastructure Leasing and Financial Services Ltd, 95
Intelligence Bureau, 203
International Herald Tribune, 69
International Monetary Fund (IMF), 43
iPhone, 110
Israel, 11–12, 14, 253n; intelligence services of, 205
izzat, 171

J. Walter Thompson (JWT), 66
Jaipur, 102
Jamaica, 11, 31, 122, 182
Japan, 11, 14, 200, 208, 219, 227n,
Java, 12
Jensen, Robert, 119, 122, 144
Jews, 11
Jharkhand, 127, 220
jihad, 22
Journal of International Development, 115

kabaadiwala, 214
Kalahari Desert, 168

INDEX

Kalighat paintings, 190
Kaliyug, 190–1
Kannada, 139, 263n
Kanshi Ram, 147–8, 150, 152–3, 158
Kanyakumari, 210
Karbonn, 80, 102
Kargil, 205
Karnataka, 196, 204, 256n
Karunanidhi, K., 61
Kashmir, 205, 210
Katz, James, 10
Kaveri River, 37
keitai, 219
Kenya, 123, 129
Kerala, xii, 22–3, 95–6, 118–19, 120–3, 132, 144, 196
King, Rodney, 159, 161
Kipling, Rudyard, 204
Kochi, 96, 118
Kolkata, 24, 26, 43, 52, 55–6, 85, 161
kotwal, 37
Kumbakonam, 21
Kuwait, 54
Kyunki Saas Bhi Kabhi Bahu Thi, 173

La Demoiselle du téléphone, 221
Lal, Dinesh, 188
Lamb, Sarah, 175
languages in India: number of, 139, 160, 194, 217, 219–20
Lava, 191
Leacock, Stephen, 125
Leone, Sunny, 191
LG, 80–1, 83, 100, 103–04
Liebling, A. J., 24
Ling, Rich, 10, 200, 215
literacy, xi, 8, 19–20, 29, 143, 150, 172, 199, *see also* illiteracy, mobile literacy

London School of Economics, 145
London, 212
Los Angeles Times, 194
Los Angeles, 159
'love marriages', 168
Love, Sex aur Dhoka, 194
Lucknow, 21, 81, 96–8, 100, 150, 154
Ludhiana, 53

machinery: 'like a snake-hole', 26
Madon, Shirin, 9
Madrid train bombing, 145, 205
Magic pre-paid card, 72–3; *see also* pre-paid cards and services
Mahanagar Telecom Nigam Ltd (MTNL), 33
Maharashtra, 71, 138–9, 201
Mahatma Gandhi National Rural Employment Guarantee Act (MNREGA), 102, 125–6, 249n
Mahe, 23
Malaysia, 11, 56
Mallah, 176, 181
mallahis, 132
Manmohan Singh, 57–8
Manucci, Niccolao, 20, 37
Mao Zedong, 153, 160
Marathas, 22, 122
marketing, 75–81
Martin, Michele, 168
Marwar, 55
Marx, Karl, 93–4
Maxis Communications, 56
Mayawati, 148–50, 152–54, 156
Mazzarrella, William, 70
McDonald's, 108
McGilvray, Dennis, 12
McLuhan, Marshall, 141
Mecca, 12
Meerut, 24

INDEX

Mehta, Vinod, 192
Mexico, 3
MGNREGA, *see* Mahatma Gandhi National Rural Employment Guarantee Act
Micro and Small Enterprises (MSE), 122
Micromax, 80
Microsoft, 211
microwave, 168
middlemen, 115, 120–2, 245n
Miller, Daniel, 10, 31, 122, 182
Miller, Eric, 23
Mirs, 12
Mishra, Satish, 152–3, 155
Misra, Pankaj, 53
mistriis, 97–101, 104, 106–09, 242n
Mittal, Sunil, 53–4, 56, 61
MMS (Multi-Media Messaging Service), 194–6, 207
Mo Ibrahim, 3
mobile literacy, 139, 141, 181, 218
Mobile Money Transfer (MMT), 127, 130
'Mobile Phone': short story, 223
mobile phones, numbers of, 6; kosher, 12; morality and 11–13, 167, 187, 190–6; religion and, 11–14, 190; *see also* teledensity, telephones, cell phones
Mobile Wali Dhobinaya, 188, 190
Mobile Wali, 187–9
mobility, 20, 21
Modi: business family, 55
Modi, Narendra, 155
monsoon: fishing during, 120
Monty Python, 25
moral economy: of boatmen, 133, 137
Morozov, Evgeny, 160, 221
Motorola, 103, 123, 128–9, 131
Mahanagar Telecom Nigam Ltd (MTNL), 33, 42–3, 50, 56, 68

Mughals, 20, 21, 23, 25, 37, 166
Mukund, A. 72
Mulayam Singh Yadav, 153, 158
Mumbai Stock Exchange, 54
Mumbai terrorist attack, 2008, 23, 117, 197, 206
Mumbai, 4, 5, 23, 26, 43, 52, 55–6, 62, 68, 76, 95, 101, 192, 196, 201, 207, 209
Musharraf, Pervez, 205
'Music Mania in Small-town Bihar', 217
Muslims, 11, 12, 78, 97, 99, 118, 214

Nair, Sumitra, 86, 98, 100
Nambudiri Brahmins, 23; *see also* Brahmins
Narasimha Rao, P. V., 43, 47
Narayan, R. K., 21
Narendra Modi, 155
Nariman House, 207
Nasik, 131
National Telecom Policy, 1994 (NTP-94), 35–7, 44
Nehru, Jawaharlal, 28, 67
Net Delusion, 160
Network Power, 162, 217
Networked, 162
networks: power of, 162
networks, social, 6, 9–11, 26, 98, 134–5, 149, 153–4, 158–9, 162, 175, 209, 211, 215–17
New Delhi, 15, 39, 55–6, 76, 101, 123, 127–8, 174
New Telecom Policy, 1999 (NTP-99), 41, 49–50
New York Times, 39, 118–119, 207
New York, 212
New Zealand, 13
newspapers, 28–9; daily, 220; ownership of, 154

INDEX

NGOs, 4, 92, 159, 213–14, 241n
Nokia Care, 103–10, 214
Nokia, 1, 40, 65, 76–8, 82–6, 90–1, 96, 99–105, 131, 174, 178, 191, 219, 223
North Africa, 216
North America, 10, 139, 208
Norway, 58

O'Neill, Peter, 115
Obama, President Barak, 149, 151
OBC, *see* Other Backward Classes
Oberoi Hotel, 206
Ogilvy and Mather, 68
Olivelle, Patrick, 143
Ontario, 125
Open Magazine, 192
Orissa, 167
Oriya, 139
Orthodox Jewish centre, 207
Other Backward Classes (OBC), 30, 78
Outlook, 70, 207

Pai, Sudha, 22
Pakistan, 116, 205–07
Palestine, 205
Panaji, 116
Panini Keypad, 219
Panipat, 22, 122
Patna, 166
Patra, Sabyasachi, 91
PCOs, *see* Public Call Offices
Pennsylvania, 168
Penthouse, 192
Perpetual Contact, 10
Peshawar, 22
Philippines, 145, 219, 221
phone tapping, 204
phones, number of, 73, 103, 151; *see also* teledensity, mobile phones

phones, social pleasures of, 139
Phony Equality, 92
pilgrims, 25, 131–5, 180
Pitroda, Sam 19, 33–4
Playboy, 192
pneumonia, 211
polling booths, in UP, 152, 155, 221
Pondicherry, 23
Pope, The, 160
pornography, 8, 13, 186, 191–7
post office: items carried, 29
postcard: cost of, 29
Prahalad, C. K., 119–20
pregnancy, 211
pre-paid cards and services, 5–6, 71–2, 74, 79, 82, 124, 127–8, 199, 205, 223
Press Trust of India, 39
priests, 21–2, 120, 134, 136, 189, 190
Prime Ministers: from lower castes, 31
printing, 20, 23–5
private investment in telecom, 48
privatization, 49–50
Public Call Offices (PCO), x, 33–4, 36, 42, 170–1
Pune, 21–2, 131
Punia, P. L., 156–9
Punjab University, 199
Punjab, 53, 212
Putin, Vladimir, 159

rabbis, 12
Radhika, A. Neela, 73
Radia, Niira, 202–03
radiation, electromagnetic, 212
Radio Frequency (RF) spectrum, xxxi-xxxii, 5–6, 36–7, 40, 45, 52, 57–8, 60–1, 65, 73, 89, 95, 142, 203, 210, 237n

INDEX

radio, 4, 29
Raebareli, 93
Rafael, Vicente, 144
Rahman, A. R., 66
railways, 4, 24–6, 28, 30, 57, 123, 216, 231n
Rainie, Lee, 162–3, 204
Raj Ghat, 132–3
Raja, A., 57, 59, 202–03
Ram, Kanshi, *see* Kanshi Ram
Ram, Sukh, *see* Sukh Ram
Ramnagar, 176
Rangaswamy, Nimmi, 86, 98, 100
Ranjit Singh, Emperor, 20, 22
recycling, 213; *see also* e-waste
Reddy family, 56
Reddy, Dr Pratap, 56
Reddy, Suneeta, 56
Rediffusion, 66
Reliance mobiles, 79–80
Reliance Telecom Limited (RTL), 70–1
Reliance, 51–2, 54–5, 74–5, 83, 178, 223; *see also* Ambani
religion: mobiles and, 11–14, 190; *see also* mobile phones, morality and;
reservation, 30, 148
Reuters, 4
Revenue Intelligence, Department of, 203
revolver, 162, 218
Rheingold, Howard, 144–5
Rise of the Network Society, The, 9
roads, 22, 37, 154, 224
Roh Moo-Hyun, 144
Roman alphabet, 219
Rosy the Riveter, 107
Ruia family, 56
Ruia, Ravi, 55
Ruia, Shashi, 55
Russia, 159

Saharanpur, 71, 104
Samajwadi Party, 156–7
Samsung, 15, 76–7, 80–1, 83, 100, 102
Sanskrit, 219
Satish Babu, D., 75–7
Sawai, Akshay, 192
State Bank of India, 131
Scheduled Castes (SCs), 30, 157 *see also* Dalits
Scheduled Tribes (ST), 30
scripts, Indian, 219; *see also* languages
Seattle, 144
Sebastian, Bobby, 73, 95–6
Second World War, 107
Seelampur, 214
Senators, US, 57
Sevagram, 26
'sexting', 194
Shah Jahan, Emperor, 21
Sharma, Ranjan, 141
Shimla, 26, 204
Shirky, Clay, 9, 197
Shiva, 180–2
shoes, 2, 105, 152, 185, 225n
Shoghi, 204
shopkeepers, 71–2, 79; EKO and, 128
Sibal, Kapil, 59
Siddiqui, Saif, 97–8
Siddiqui, Salim, 97
SIM card, 80, 83, 85, 99, 117, 127, 140, 205; *see also* Subscriber Identity Module
Singapore, 54
Singh, Manmohan, *see* Manmohan Singh
Singtel, 54
Sinha, Abhinav, 123–4
Sinha, Abhishek, 123–4, 131
Skycell, 75

INDEX

Slim, Carlos, 3
small business, 116; *see also* shopkeepers
Smart Mobs, 144
SMS, 13, 82, 138, 153–4, 160, 166, 185, 210, 218–19
sociability, 10
Sony Ericsson, 80–2
Sony, 76–7, 80–1, 105, 108–09
sousveillance, 159–60
South Africa, 14, 158–9
South Korea, 144
Spanish general elections, 2004, 144
Special Economic Zones (SEZ), 40, 90–1, 93
Spice, 55, 102
Spider Man software, 102
spying, 13, 19, 205, 229n; *see also* surveillance
Sreekumar, T. T., 121
Sri Lanka, 12, 54
Sriperumbudur, 39–40, 90–1, 103
Srivastava, Sanjay, 195
Stalin, Josef, 160
Star Global Resources Ltd, 140
Star Network, 69
State Bank of India (SBI), 124
State Domestic Product (SDP), 115
Subscriber Identity Module (SIM), 73–4; *see also* SIM
subscribers, 6, 7, 73; *see also* phones: numbers of, teledensity
Sudra, 22, 143
suicide, 9, 39, 196, 247n
Sukh Ram, 45–7, 61
Sullivan, Kevin, 118, 120
Supreme Court, 52, 60, 89, 186, 203, 206
Surat, 21

surveillance, 13, 141, 202
Susan, Nisha, 195
Swami, Praveen, 204–05
Swamy's Treatise on Telephone Rules, 42, 47, 65
Sweden, 159
Sydney, 31

Taj Mahal Hotel, 206
Tamil Nadu, 21, 56, 61, 73, 90–1, 202, 222
tapping, phone, 204; *see also* spying, surveillance
Tata group, 51, 55–6, 202, 223
Tata Telecommunications, 55
Tata Teleservices, 74
Tehelka, 8, 195, 212
Telecom Dispute Settlement and Appellate Tribunal (TDSAT), 50, 52
Telecom Regulatory Authority of India (TRAI), 41, 43, 47–8, 50, 52, 59, 196–7
Telecom Restructuring Committee, 43
Telecommunications, Department of, 33, 47–8, 50–1, 58
teledensity, 6, 28, 34, 41, 44, 48, 50, 67; in 1984, 32; *see also* subscribers
telegraph, 4, 25–6
Telekom International Malaysia, 55
Telenor, 58
telephone: introduction of, 26; not a priority in India, 32; number of, 27–8; *see* teledensity
television, 29–30, 168, 197, 205, 214, 222, 238n
Tendulkar, Sachin, 66
Tenhunen, Sirpa, 169
terrorism, 205–07

291

INDEX

terrorists, 8, 102, 142, 146, 197, 202
text messages, 218–19; in Indian languages, 139; *see also* SMS
thalidomide, 212
The Wire, 205
theft: of phones, 198–9
Thomas, Thomas K., 58
Tihar Jail, 61
Times of India, 27, 166, 200
Tiwari, Manoj, 187–9
Toronto, 144
tourists, 131–6
towers: transmission, xxxii, 6, 57, 62, 72, 74, 93–7, 119, 142, 160, 211–13; *see also* health
Toxic Link, 214
training: sales people, 83–5
Travancore, 26
Trinidad, 182
Tripathy, Ratnakar, 217
turnout: voters in UP, 155
TVS Group, 95
two-step flow, 141

UFS3 software, 102
Uganda, 169
UK, 35
Unique Identification Authority of India (UIAI), 222
Unitech, 58
United Arab Emirates, 11
United Nations Environmental Program (UNEP), 214
United Nations, 35
UniverCell, 75–8
Universal Access Service Licence (UASL), 68
Unlawful Activities Prevention Act (UAPA), 207
untouchability, 30
Untouchables, 145

Uttar Pradesh (UP), 8, 71, 77, 84, 91, 103, 146–7, 151–2, 154–5, 158, 163, 167, 193, 200, 220, 222; elections in, 147–9
USA, 35, 51–2, 56–7, 67, 111, 139, 147, 196, 200

Vajpayee, Atul Bihari, 41, 49, 155, 235n
Varanasi, *see* Banaras
Varghese, Anupam, 125–7
Vedas, Vedic literature, 22, 143
Veeraraghavan, Rajesh, 138
Vernacular Press Act, 1878, 24
Vice-President of USA, 57
Videsh Sanchar Nigam Ltd (VSNL), 33
Vodafone, 55, 69, 71, 73, 75, 94, 131, 211, 216, 223

waiting list: for telephone, 41
Walton, Marion, 158
Washington Post, 120
watch: pocket and wrist, 2, 10, 117, 136, 177
Welcome to Sajjanpur, 29
Wellman, Barry, 134, 162–3, 204, 215–16
West Africa, 5
West Asia, 35
West Bengal, 169
White, E. B., 100
White, Richard, 57
Wikipedia, 69
Wolff, 4
women, 11, 19, 25, 82–3, 111, 129–31, 165–83, 190–1, 207, 211, 216; absent among *mistriis*, 101; bank accounts and, 130; employment in Nokia factory, 90–4, 103, 241n; Nokia repair

centre and, 106–7; music videos and, 187–9
World Bank, 3, 43, 211
World Health Organisation, 212
World Trade Organisation, 144
writing, 20; *see also* literacy, illiteracy

Yadav, Akhilesh, 157
Yadav, Mulayam Singh, *see* Mulayam Singh Yadav
Yadavs, 78
YouTube, 92, 156, 187

Zain, 54